David King
603-885-6717

Improving Product Reliability

Wiley Series in Quality and Reliability Engineering

Editor
Patrick D.T. O'Connor
www.pat-oconnor.co.uk

Electronic Component Reliability:
Fundamentals, Modelling, Evaluation and Assurance
Finn Jensen

Integrated Circuit Failure Analysis:
A Guide to Preparation Techniques
Friedrich Beck

Measurement & Calibration Requirements
For Quality Assurance to ISO 9000
Alan S. Morris

Accelerated Reliability Engineering:
HALT and HASS
Gregg K. Hobbs

Test Engineering:
A Concise Guide to Cost-effective Design, Development
and Manufacture
Patrick D.T. O'Connor

Improving Product Reliability:
Strategies and Implementation
Mark Levin and Ted Kalal

Improving Product Reliability

Strategies and Implementation

Mark A. Levin and Ted T. Kalal

Teradyne, Inc., California, USA

WILEY

Other Wiley Editorial Offices

John Wiley & Sons Inc., 111 River Street, Hoboken, NJ 07030, USA

Jossey-Bass, 989 Market Street, San Francisco, CA 94103-1741, USA

Wiley-VCH Verlag GmbH, Boschstr. 12, D-69469 Weinheim, Germany

John Wiley & Sons Australia Ltd, 33 Park Road, Milton, Queensland 4064, Australia

John Wiley & Sons (Asia) Pte Ltd, 2 Clementi Loop #02-01, Jin Xing Distripark, Singapore
129809

John Wiley & Sons Canada Ltd, 22 Worcester Road, Etobicoke, Ontario, Canada M9W 1L1

Wiley also publishes its books in a variety of electronic formats. Some content that appears
in print may not be available in electronic books.

British Library Cataloguing in Publication Data

A catalogue record for this book is available from the British Library

ISBN 0-470-85449-9

Typeset in 10.5/13pt Sabon by Laserwords Private Limited, Chennai, India
Printed and bound in Great Britain by TJ International, Padstow, Cornwall
This book is printed on acid-free paper responsibly manufactured from sustainable forestry
in which at least two trees are planted for each one used for paper production.

Cary and Darren Kalal

To my beautiful wife, Dana Mischel Levin,
for her endless love, support, and patience, and to our sons,
Spencer and Andrew.

Contents

About the Authors

Ted Kalal is a practicing reliability engineer who has gained much of his under-standing of reliability from hands-on experience and from many great mentors. He is a graduate of the University of Wisconsin (1981) in Business Admin-istration after completing much preliminary study in mathematics, physics, and electronics. He has held many positions as a contract engineer and as a consultant, where he was able to focus on design, quality, and reliability tasks. He has authored several papers on electronic circuitry and holds a patent in the field of power electronics. With two partners, he started a small manufacturing company that makes hi-tech power supplies and other scientific apparatus for the bioresearch community. In 1999, he joined Teradyne, Inc. as a Reliability Manager. (reliability2help@aol.com)

Mark Levin is a senior reliability design engineer at Teradyne, Inc. in Agoura Hills, California. He received his Bachelor of Science degree in Electrical Engineering (1982) from the University of Arizona and a Masters of Science degree in Technology Management (1999) from Pepperdine University. He has more than 20 years of electronics experience spanning the aerospace, defense, consumer, and medical electronics industries. He has held several manage-ment and research positions at Hughes Aircraft's Missiles Systems Group and Microwave Products Division, General Medical Company and Medical Data Electronics. His experience is diverse, having worked in manufacturing, design, and research and development. He has developed manufacturing and reliability design guidelines, reliability training, workmanship standards, quality pro-grams, JIT manufacturing, ESD safe work environments, and installed surface mount production capability. (reliabilityaid@aol.com)

Series Foreword

Modern engineering products, from individual components to large systems, must be designed and manufactured to be reliable in use. The manufacturing processes must be performed correctly and with the minimum of variation. All of these aspects impact upon the costs of design, development, manufacture, and use, or, as they are often called, the product's life cycle costs. The challenge of modern competitive engineering is to ensure that life cycle costs are minimized whilst achieving requirements for performance and time to market. If the market for the product is competitive, improved quality and reliability can generate very strong competitive advantages. We have seen the results of this in the way that many products, particularly Japanese cars, machine tools, earthmoving equipment, electronic components, and consumer electronic products have won dominant positions in world markets in the last 30 to 40 years. Their success has been largely the result of the teaching of the late W. E. Deming, who taught the fundamental connections between quality, productivity, and competitiveness. Today this message is well understood by nearly all the engineering companies that face the new competition, and those that do not understand lose position or fail.

The customers for major systems, particularly the US military, drove the quality and reliability methods that were developed in the West. They reacted to a perceived low achievement by the imposition of standards and procedures, whilst their suppliers saw little motivation to improve, since they were paid for spares and repairs. The methods included formal systems for quality and reliability management (MIL-Q-9858 and MIL-STD-758) and methods for predicting and measuring reliability (MIL-STD-721, MIL-HDBK-217, MILSTD-781). MIL-Q-9858 was the model for the international standard on quality systems (ISO9000), and the methods for quantifying reliability have been similarly developed and applied to other types of products and have been incorporated into other standards such as ISO60300. These approaches have not proved to be effective, and their application has been controversial.

By contrast, the Japanese quality movement was led by an industry that learned how quality provided the key to greatly increased productivity and competitiveness, principally in commercial and consumer markets. The methods that they applied were based upon an understanding of the causes of variation and failures, and continuous improvements through the application of process controls and motivation and management of people at work. It is one of history's ironies that the foremost teachers of these ideas were Americans, notably P. Drucker, W.A. Shewhart, W.E. Deming and J.R Juran.

These two streams of development epitomize the difference between the deductive mentality applied by the Japanese to industry in general, and to engineering in particular, in contrast to the more inductive approach that is typically applied in the West. The deductive approach seeks to generate continuous improvements across a broad front, and new ideas are subjected to careful evaluation. The inductive approach leads to inventions and 'break-throughs', and to greater reliance on "systems" for control of people and processes. The deductive approach allows a clearer view, particularly in discriminating between sense and nonsense. However, it is not as conducive to the development of radical new ideas. Obviously these traits are not exclusive, and most engineering work involves elements of both. However, the overall tendency of Japanese thinking shows in their enthusiasm and success in industrial teamwork and in the way that they have adopted the philosophies of western teachers such as Drucker and Deming, whilst their western competitors have found it more difficult to break away from the mould of 'scientific' management, with its reliance on systems and more rigid organizations and procedures.

Unfortunately, the development of quality and reliability engineering has been afflicted with more nonsense than any other branch of engineering. This has been the result of the development of methods and systems for analysis and control that contravene the deductive logic that quality and reliability are achieved by knowledge, attention to detail, and continuous improvement on the part of the people involved. Therefore it can be difficult for students, teachers, engineers, and managers to discriminate effectively, and many have been led down wrong paths.

In this series we will attempt to provide a balanced and practical source covering all aspects of quality and reliability engineering and management, related to present and future conditions, and to the range of new scientific and engineering developments that will shape future products. The goal of this series is to present practical, cost efficient and effective quality and reliability engineering methods and systems.

I hope that the series will make a positive contribution to the teaching and the practice of engineering.

Patrick D.T. O'Connor
February 2003

Foreword

In my 26 years at Teradyne, I have seen the automated test industry emerge from its infancy and grow into a multi-billion-dollar industry. During that period, Teradyne evolved into the world's leading supplier of automated test equipment (ATE) for testing semiconductors, circuit boards, modules, voice, and broadband telephone networks. As our business grew, the technology necessary to design ATE became increasingly complex, often requiring leading-edge electronics to meet customer performance needs. Our designs have pushed the envelope, demanding advancements in nearly every technological area including process capability, component density, cooling technology, ASIC complexity, and analog/digital signal accuracy.

Our customers, too, insist on the highest performance systems possible to test their products. But performance alone does not provide the product differentiation that wins sales. Customers also demand incomparable reliability. Revenue lost when an ATE system goes down can be staggering, often in the tens of thousands of dollars per hour. Furthermore, because of design complexity and system cost, the warranty cost to maintain these systems is increasing. Low reliability severely impacts the bottom line and impedes the ability to gain and hold market share.

To improve product reliability, changes had to be made to the reliability process. We learned that the process needed to be proactive. It had to start early in the product concept stage and include all phases of the product development cycle. In researching solutions for improving product reliability, we found the wealth of information available to be too theoretical and mathematically based. Clearly, we didn't want a solution that could only be implemented by reliability engineers and statisticians. If the training were overly statistical, the message would be lost. If the process required training everyone to become a reliability engineer, it would be useless. The process had to reduce technical reliability theory into practical processes easily understood by the product development team.

For the reliability program to be successful, we needed a way to provide both management and engineering with practical tools that are easily applied to the product development process. The reliability processes presented in this book achieves this goal.

The authors logically present the reliability processes and deliverables for each phase of the product development cycle. The reliability theory is thoughtful, easily grasped, and does not include a complex mathematical basis. Instead, concepts are described using simple analogies and practically based processes that a competent product development team can understand and apply. Thus, the reliability process described can be implemented into any electronic or other business regardless of its size or type, and ultimately helps give customers products with superior performance and superior reliability.

Edward Rogas, Jr.
Senior Vice President
Teradyne, Inc.

Preface

Nearly everyday, we learn of another company that has failed. In the new millennium, this rate of failure will increase. Competitors are rapidly entering the market place using technology, innovation, and reliability as their weapons to gain market share. Profit margins are shrinking. Internet shopping challenges the conventional business model. The information highway is changing the way consumers make buying decisions. Consumers have more resources available for product information, bringing them new awareness about product reliability.

These changes have made it easier for consumers to choose the best product for their individual needs. As better-informed shoppers, consumers can now determine their product needs at any place, anytime, and for the best price. The information age allows today's consumer to research an entire market efficiently at any time and with little effort. Conventional shopping is being replaced by "smart" shopping. And a big part of smart shopping is getting the best product for the best price.

As the sources for product information continue to increase, the information available about the quality of the product increases as well. In the past, information on product quality was available through consumer magazines, newspapers, and television. The information was not always current and often did not cover the full breadth of the market. Today's consumer is using global information sources and Internet chat to help in their product-selection process. An important part of the consumer's selection process is information regarding a product's quality and reliability. Does it really do what the manufacturer claims? Is it easy to use? Is it safe? Will it meet customer expectations of trouble-free use? The list can be very long and very specific to the individual consumer.

From automobiles to consumer electronics, the list of manufacturers who make high-quality products is continuously evolving. Manufacturers who did not participate in the quality revolution of the last two decades were replaced by those that did. They went out of business because the companies with high-quality systems were producing products at a lower cost. Today, consumers demand products that not only meet their individual needs but also meet these needs over time. Quality design and manufacturing was the benchmark in the 1980s and 1990s; quality over time (reliability) is becoming the requirement in the twenty-first century. In today's marketplace, product quality is necessary in order to stay in business. In tomorrow's marketplace, reliability will be the norm.

Quality and reliability are terms that are often used interchangeably. While strongly connected, they are not the same. In the simplest terms:

- Quality is conformance to specifications.
- Reliability is conformance to specification *over time.*

As an example, consider the quality and reliability in the color of a shirt. In solid color men's shirts, the color of the sleeves must match the color of the cuffs. They must match so closely that it appears that the material came from the same bolt of cloth. In today's manufacturing processes, several operations occur simultaneously. One bolt of cloth cannot serve several machines. The colors of several bolts of cloth must be the same, or the end product will be of poor quality. Every bolt of cloth has to match to a specified color standard, or the newest manufacturing technologies cannot be applied to the process. Quality in the material that goes into the product is as important as the quality that comes out. In fact, the quality that goes in becomes a part of the quality that comes out. After numerous washings, the shirt's color fades out. The shirt conformed to the consumer's expectations at the time of first use (quality) but failed to live up to the consumers' expectations (reliability).

Reliability is the continuation of quality over time. It is simply the time period over which a product meets the standards of quality for the period of expected use. Quality is now the standard for doing business. In today's marketplace and beyond, reliability will be the standard for doing business. The quality revolution is not over; it has just evolved into the reliability revolution.

This book is an effort to guide the user on how to implement and improve product reliability with a product life cycle process. It is written to appeal to most types of businesses regardless of size. To achieve this, the beginning of each chapter discusses issues and principles that are common to all businesses,

Table 1 Business Size Definition

Metric	Company size		
	Small	Medium	Large
Employee count	<100	>100 & <1000	>1000
Gross sales dollar	<$10 M	>$10 M & <$100 M	>$100 M
Dollars available from the warranty budget (approx.)	<$1 M	$1 M to $10 M	>$10 M

independent of size. We also segregate business into three categories based on size: Small, Medium, and Large. Definitions are summarized in Table 1.

The finance department can, more precisely, quantify the lost revenue due to warranty claims and poor quality. This loss represents the potential dollars that are recoverable "after" the reliability process improvements have been implemented and have begun to bear fruit.

GAINING COMPETITIVE ADVANTAGE

Manufacturers, who have no reliability engineering in place, typically have warranty costs as high as 10 to 12% of their gross sales dollar. A company that implements reliability into their processes can see warranty costs diminish to below 1% of the gross sales dollar. The total amount that can be recovered from the warranty budget represents the dollars that could be reinvested (from the warranty budget) or added to earnings. If research and development is 10% of the gross sales dollars, then the annual warranty dollar savings from reliability can cover the costs to develop future products. Of course, this only addresses the tangible benefits from a reliability program. There are many intangible benefits that are gained by improving product reliability. Examples include better product image, reduced time to market, lower risk of product recall and engineering changes, and more efficient utilization of employee resources. These intangible assets are addressed later in the book.

List of Acronyms

ALT	Accelerated Life Testing
ASIC	Application-specific Integrated Circuit
BOM	Bill of Materials
DDT	Device Defect Tracking
DFM	Design for Manufacturing
DFR	Design for Reliability
DFS	Design for Service (and maintainability)
DFT	Design for Test
DMT	Design Maturity Test
DOE	Design of Experiments
DVT	Device Verification Test
DUT	Device Under Test
ECO	Engineering Change Order
ESD	Electrostatic Discharge
ESS	Environmental Stress Screening
FA	Failure Analysis
FBD	Functional Block Diagram
FIFO	First In, First Out
FIT	Failures In Time
FMEA	Failure Modes and Effects Analysis
FRACAS	Failure Reporting, Analysis and Corrective Action System
FRU	Field-Replacement Unit
FTA	Fault Tree Analysis
HALT	Highly Accelerated Life Test
HASA	Highly Accelerated Stress Audit
HASS	Highly Accelerated Stress Screens
HAST	Highly Accelerated Stress Test
HTOL	High-Temperature Operating Life test
ICM	Identify, Communicate, and Mitigate
JIT	Just in Time
LN2	Liquid Nitrogen
MRB	Material Review Board
MTBF	Mean Time Between Failures
MTBM	Mean Time Between Maintenances
MTTF	Meat Time To Failure

MTTR	Mean Time To Repair
MTTRS	Mean Time To Restore System
NRE	Nonrecurring Engineering
PCB	Printed Circuit Board
POS	Proof of Screen
PPM	Parts Per Million
RDT	Reliability Demonstration Test
RPN	Risk Priority Number
SPC	Statistical Process Control
TQM	Total Quality Manufacturing
VOC	Voice of the Customer

Acknowledgements

We would like to recognize and thank Harding Ounanian for his significant contribution in doing the first edit of the book.

Special thanks also to Joel Justin, Kevin Giebel, Jim McLinn, Pat O'Connor, Steve King, Dana Levin, and Larry Steinhardt for their technical edits and Glenn Hemanes for his patience and help with some of the artwork. Finally, we would like to thank Ed Rogas for the foreword and for supporting our work.

We would also like to thank the following people who have brought a better awareness about reliability and continue to influence our way of thinking; Benton Au, Joe Denny, Dave Evans, Jim Galuska, Ray Hansen, Dr. Greg Hobbs, Jim McLinn, Pat O'Connor, Roy Porter, and Dr. David Steinberg.

Part I

Reliability – It's a Matter of Survival

1

Competing in the Twenty-first Century

Reliability, why do you need it? The major US car manufacturers saw their dominance eroded by the Japanese automobile manufacturers during the 1970s because the vehicles produced by the big three had significantly more problems. The slow downward market slide of the US automobile industry was predictable when the defect rate of US automobiles is compared with the Japanese automobile industry. In 1981, a Japanese-manufactured automobile averaged 240 defects per 100 cars. The US automobile manufacturers during the same time period were manufacturing vehicles with 280 to 360% more defects per 100 vehicles. General Motors averaged 670 defects per 100 cars, Ford averaged 740 defects per 100 cars, and Chrysler was the highest with 870 defects per 100 cars.

Much has been written about how this came about and how the US manufacturers began implementing Total Quality Manufacturing (TQM), quality circles, continuous improvement, and concurrent engineering to improve their products. Now the US automobile industry produces quality vehicles and the perception that Japanese vehicles are better has eroded significantly. J. D. Powers and Associates reported in their 1997 model year report that cars and trucks averaged about 100 defects per 100 vehicles. This represented a 22% increase from 1996 and a 100% decrease from 1987. Vehicles such as the GM Saturn and Ford Taurus are a tribute to the success both in financial terms and in improving the perception that automobile manufacturers in the United States can produce reliable, quality automobiles.

In the 1970s, the typical automobile warranty was for 12 months or 12,000 miles. In 1997, automobile manufactures were offering 3-year/36,000-mile bumper-to-bumper warranties. Three years later, these same automobile manufacturers were offering 7-year/100,000-mile warranties. Jaguar is now

Improving Product Reliability: Strategies and Implementation. Mark A. Levin and Ted T. Kalal
© 2003 John Wiley & Sons, Ltd ISBN: 0-470-85449-9

advertising a 7-year/100,000-mile warranty on its used vehicles! BMW has responded with a similar type of program. The reason these manufacturers can offer longer warranty periods is because they understand why and how their vehicles are failing and can therefore produce more reliable vehicles.

A 1997 consumer reports survey of 604,000 automobile owners showed a dramatic improvement in the perception of the reliability of US-manufactured automobiles. The improvement by the big three (automobile manufacturers) did not occur overnight. It was the result of a commitment to provide the necessary resources along with a credible plan for producing the reliable vehicles. It was a paradigm change that took years and evolved through many steps.

1.1   GAINING COMPETITIVE ADVANTAGE

Companies successfully competing in the twenty-first century will share a common thread. They will all produce quality products that meet or exceed customer expectations over time. This may not seem like new information, though the process and tools that these companies will use to achieve this will be new. In some industries, technology moves so fast that customers tend to trade up to the next-generation product before it stops performing to specification. This may seem like the ideal environment for a manufacturer because the product life expectations of the consumer are shorter. However, in reality, achieving product reliability with decreasing product development times requires a change in the way we develop products. Platform product development times have shortened to eighteen months and their derivatives (product offshoots) have shrunk to twelve months or less. Of course, this is highly dependent on the product complexity, regulatory and safety requirements, but the trend cannot be ignored. Companies pay a heavy price for releasing a product that is "buggy" or unreliable. Satisfied customers are repeat customers. It is a well-known fact that it costs between 5 to 10 times more to acquire new customers than it does to retain existing ones. It doesn't matter whether you are competing on cost or product differentiation; reliable products result in repeat customers and product growth through word of mouth. A faulty product usually results in the customer communicating dissatisfaction to anyone who will listen until the product or service is replaced with a more reliable one.

1.2   COMPETING IN THE NEXT DECADE – WINNERS WILL COMPETE ON RELIABILITY

The business practices of the past few decades will not be sufficient to ensure success in the twenty-first century. Through the years, we've learned to master the skill of building quality products. Higher quality products have resulted in improved profit margins. In fact, consumers make buying decisions based on their perception of which products had better quality when the competing products were of the same approximate price. In the past few decades, reliability was not a deciding factor for most consumers. This is mostly the result of the consumer's lack of knowledge about product quality. However, the average consumer in the twenty-first century will make buying decisions based not only on price and quality but also on the perceived reliability of the product. Consumers make buying decisions based on which product offers the best value. We can define product value as

$$\text{Product Value} = \frac{\text{Customer Perceived Value}}{\text{Price}}$$

Here the customer-perceived value is related to the quality and reliability of the product. One of the key advantages of implementing reliability throughout the organization and at every phase of the product life is that the product value increases because of an improved customer perception of the value of the product and the lower cost of production. There is a common misperception that implementing reliability delays the product development time and increases the cost of the product (both material and production costs). But the reality is the exact opposite. Products that are more reliable generally have lower production costs. The reason for this is the result of many factors that contribute to reducing product costs and the product development cycle. For example, products that are reliable generally have

- higher first pass yield in test,
- less material scrap,
- less product rework (which helps to lower product cost and improve product reliability),
- fewer field failures,
- reduced warranty costs (this saving can be passed onto the consumer to provide a competitive price advantage),
- lower risk of recall,
- better designs that are easier to manufacture.

Looking back at the definition of what the consumer considers to be of value, it becomes clear that product reliability will increase the perceived product value and lower the cost of production. This is an important fact about product reliability that is often misunderstood.

1.3   CONCURRENT ENGINEERING

An important ingredient for successful design and implementation of new technologies into manufacturing involves the establishment of concurrent engineering practices. Concurrent engineering is a process used from design concept through product development and into manufacturing in which cross-functional representatives from all relevant departments provide input on key decisions. These decisions have a direct impact on the price, performance, quality, and development time required for the product. The concept has been discussed extensively in the 1990s and has resulted in better products, shorter product development times and greater profits for those who use it. However, the cross-functional teams consisted of marketing, design, test, and manufacturing. The teams did not include a separate representative for reliability since this was considered part of the design and test engineer's responsibility. (We will see in Part 2 of this book that the tools used to improve product reliability are unknown to most design and test engineers.)

This convention needs to be changed and a more encompassing version of concurrent engineering developed that takes into account the entire product life cycle. The product life cycle approach includes reliability, serviceability, and maintainability inputs that begin in the design concept phase and continue through product development and product life. This cradle-to-grave approach ensures that the lessons learned along the way are captured and incorporated into the next development cycle. Previous approaches to product development relied heavily on early Design For Manufacturing (DFM) effort and prototype testing to catch design flaws prior to product release. The problem with this approach is that DFM engineers (being highly skilled in the manufacturing process) primarily ensure that a product is manufacturable and can be rapidly ramped in production to meet market forecasts. Put another way, DFM ensures that the products designed can be ramped in production with ease (high-quality products) but the effort contributes little to product reliability.

Testing performed at the prototype stage will validate product performance to specification prior to engineering release. This does not, however, consider the ability of every product produced to meet specifications in manufacturing. The problem with this approach is that decisions are continuously made in product development that have significant impact on the product performance, the reliability, and ease with which the product can be serviced and maintained. At this stage, decisions need to be made fast. They include inputs from everyone affected, that is, marketing, design, test, manufacturing, field service, and reliability.

As stated earlier, it is important to involve all the relevant organizations and support groups early in the product development cycle in order to ensure the lowest product cost and highest product reliability. Programs such as Design For Manufacturing (DFM), Design For Test (DFT), Design For Reliability (DFR), Design for Service (DFS) and maintainability must be considered early in the product concept phase. Representatives of each of these functions provide inputs based on guidelines developed from industry standards, lessons learned, intellectual property, and internal process development. These decisions must be made on the basis of facts, not perceptions.

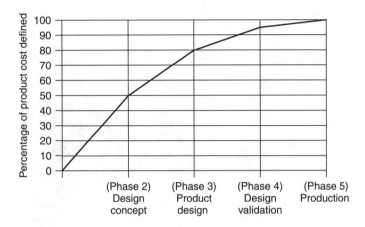

Figure 1.1   Product cost is determined early in development

Applying these design guidelines concurrently to product development will continuously reduce cost and cycle time and also optimize reliability. Figure 1.1 illustrates how a product life cycle approach to product development will have a direct, positive impact on the cost of the product. Typically, 80% of product cost is committed by the time it goes into prototyping. *Consequently, the greatest opportunity to reduce the cost of a product is in the design phase.* The product life cycle approach addresses all issues that affect the cost, service, reliability, and maintainability of the product for the entire life cycle of the product. These activities include involvement of the entire team on decisions that affect new technologies, packages, processes, and designs, and are based on a cost–benefit analysis, which includes market research risk and reliability.

Another driving reason for incorporating reliability as early as possible into product development is the cost of a change based on manpower and capital, when it is made after the design concept phase. Figure 1.1 illustrates how dramatic this impact can be on product cost. The greatest opportunity to cost is in the development and design concept phase where risk issues relating to technology, components, and processes determine the majority of the product cost. By applying these practices early in the design phase, the cost and labor resources required for implementing engineering changes can be greatly reduced.

This point is illustrated further in Figure 1.2 in which the total cost of an engineering change can increase by several orders of magnitude when it is made late in the product development cycle.

1.4   REDUCING THE NUMBER OF ENGINEERING CHANGE ORDERS (ECOs) AT PRODUCT RELEASE

In Part 4, "Reliability Process for Product Development", we will show how using tools such as Highly Accelerated Life Test (HALT)™, Highly Accelerated

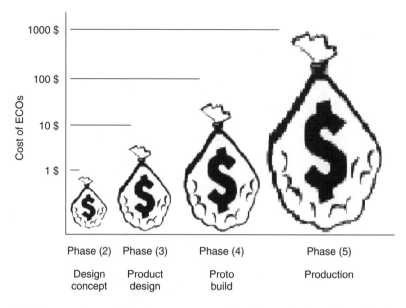

Figure 1.2 Cost to fix a design increases an order of magnitude with each subsequent phase. Courtesy of Teradyne, Inc.

Stress Screening (HASS)™, Failure Modes and Effects Analysis (FMEA) and risk mitigation early in the product development cycle will reveal hidden problems that are usually not caught until the product has been in production for some time. The product life cycle approach will also reduce the number of engineering changes at product introduction and increase long-term product reliability. This idea is best illustrated in Figure 1.3 in which a product life cycle approach ensures that the majority of the engineering design changes occur early in the development cycle. This is the best way to reduce the risk of field returns after product release. The graph illustrates how the number of field returns and engineering changes is significantly reduced through early implementation of reliability in the product development cycle.

1.5   TIME-TO-MARKET ADVANTAGE

One of the driving forces affecting product reliability is in greatly reducing the product development cycles that organizations are facing. Coincidentally, this is also the biggest reason a product cannot undergo the additional activities required for reliability. But the argument is contrary to what actually happens when reliability is included early in the design concept phase. A major advantage to the implementation of a product life cycle approach to reliability is the reduced development time for product introduction. When time-to-market goals are achieved, the benefits include product name recognition, the ability to

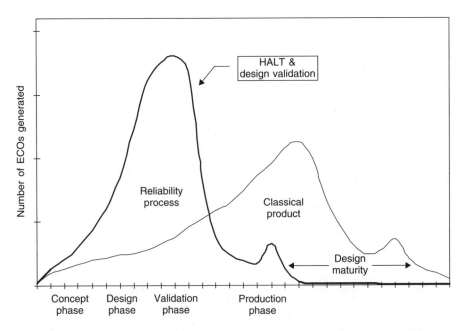

Figure 1.3 The reliability process reduces the number of ECOs required after product release

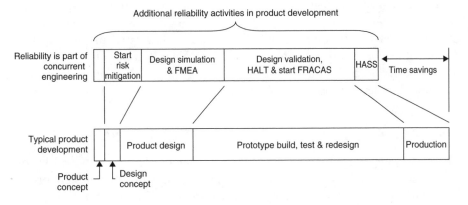

Figure 1.4 Including reliability in concurrent engineering reduces time to market

set industry standards, recognition as a leader, expansion of the customer base, and the maximization of profits. Using a product life cycle approach, product development time will be significantly reduced, as is shown in Figure 1.4.

Finally, Figure 1.5 illustrates how the timing of product introduction can affect product profitability. Introducing a new product at the same time as the competition will lead to average profits over the life of the product. By releasing a product ahead of the competition, the opportunity for profits increases.

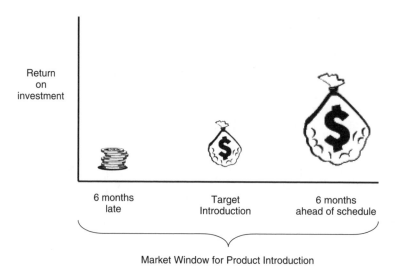

Figure 1.5 Product introduction relative to competitors. Reprinted with the permission of The Free Press, a Division of Simon & Schuster Adult Publishing Group, from REVOLUTIONIZING PRODUCT DEVELOPMENT: Quantum Leaps in Speed, Efficiency, and Quality by Steven C. Wheelwright and Kim B. Clark. Copyright 1992 by Steven C. Wheelwright and Kim B. Clark

Conversely, when releasing a product after the competition, the opportunity for profits is much lower. It is important to point out that getting too far ahead of the market can be undesirable. This point is illustrated in the article, "*A Survey of Major Approaches for Accelerating New Product Development*" by M. Millson, S.P. Raj, and D. Wilemon pg. 55, in which a late entrant in the memory chip market may not receive any profit or even recoup its investment.

1.6 ACCELERATING PRODUCT DEVELOPMENT

There are many ways to accelerate the development of a new product. Product development can be accelerated by simplifying the product design process, improving communications between the cross-functional organizations, implementing an escalation process to resolve conflicts, eliminating unnecessary steps, maintaining development workload at no higher than 85% of capacity, parallel processing as much as possible and most importantly, eliminating delays. In today's competitive technology environment, companies can no longer afford to be late with a new product release, especially in a market in which product life continues to decrease. It is important for a company to eliminate the "not invented here" attitude that can often lead to overlooking a simpler solution or a new opportunity. Another way to simplify manufacturing and to shorten the product development cycle is by standardizing common designs with a modular structure. Hewlett Packard has been very successful in creating products with modular design. Modularity renders itself to easy upgrades and new design features. In addition, by standardizing certain common features inherent to all products, the development time is greatly reduced. This also eliminates the problem of having many different versions of a common feature. For example, a common circuit like an amplifier with 10 dB of gain could be designed differently by each engineer but would perform the same function. If the 10-dB amplifier was standardized, then engineering time would not be wasted on "reinventing the wheel" and there would be a high assurance that the new circuit would work.

1.7   IDENTIFYING AND MANAGING RISKS

The key to any reliability program is the identification of risk. This concept is addressed in great detail in Part 4 in which the tools and process for implementing reliability into the product life cycle process are presented. A credible reliability program must focus on the high-risk issues in the project. There will be risk issues at every stage of the product life cycle. Early in the concept phase, decisions are made relating to the features and specifications needed to capture the target market. Marketing uses extensive Voice Of the Customer (VOC) to identify the next high-growth opportunity. These growth opportunities usually involve new technologies. For businesses that compete on the cutting edge of technology, new technologies represent a significant portion of the risk to the program and long-term reliability.

To develop the new platform product, a list of challenges must be devised. Each of these challenges represents risk to the program. To manage the risk, each item should be ranked on the severity of the risk and those items representing the highest risk should be tracked through the program. The role of reliability in the concept phase is to ensure that the risk issues are properly identified. They should be ranked by severity with corrective actions listed so that when completed the risk is mitigated. The risk plan needs to include all the functions that are affected in the life cycle of the product. Risk issues that relate to maintenance, manufacturability, design, safety, and environment are included. Unfortunately, these activities are reactive and thus add to the program development time. However, it will be shown later that the net result of these activities is the reduction in product development cost.

There are also proactive reliability activities that occur in advance of product development that help reduce product development time and improve product reliability. An example of a proactive reliability activity to reduce technology risk is the technology road map. Early VOC will identify future market needs that require new technologies. Mitigating technology risk issues in advance of need provides the required time to fully mitigate all risk issues. By mitigating technology risk in advance of need, you can gain competitive advantage by being the first to market, capturing a greater market share than would otherwise have been possible.

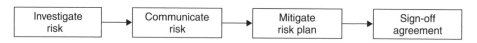

Figure 1.6 The ICM process

1.8 ICM, A PROCESS TO MITIGATE RISK

One of the biggest challenges in any development program is to identify and mitigate the risk issues early in the program. This can be best achieved through a technique called Identify, Communicate, and Mitigate (ICM) that is illustrated in Figure 1.6. The ICM approach is a three-step process to identify significant risk to the program, to communicate its impact and to develop an agreed-upon strategy to mitigate the risk. The ICM approach is an effective way to allocate and utilize limited resources in a way that will most benefit the program goals.

To identify the risk, each functional group reviews the product concepts, designs, and processes. Each group then identifies issues in which the present technology, methods, materials, processes, and tools cannot ensure success. These are the risks. Early risk identification makes the entire team focus on the concept of product reliability at the earliest possible time. All risk issues need to be identified – no matter how small – during this phase of the process. The low-risk issues will not have the same visibility and priority as the high-risk issues. In order to ensure that the major risk issues get completed prior to first customer ship, all risk issues are captured at the earliest possible point, the severity is ranked and a risk mitigation plan is put in place based on the severity of each issue. Risk identification is a process that must be formalized and documented with the following captured information:

1. Description of the risk
2. Brief description of the activity needed to mitigate risk
3. Impact of risk to program and other functional groups
4. Severity of the risk defined on a scale from insignificant to catastrophic
5. Alternate solutions.

Reference

Concurrent engineering:

1. D. R. Hoffman, *An Overview of Concurrent Engineering*, 1997 Proceedings Annual Reliability and Maintainability Symposium and 1998 Proceedings Annual Reliability and Maintainability Symposium (1998).

2

Barriers to Implementing Reliability

2.1 LACK OF UNDERSTANDING

Probably the greatest barrier to improving reliability is the understanding of what reliability actually is. Much of the resistance you will experience in implementing reliability will come from individuals who believe that quality and reliability are the same thing. But as we have shown before, they are not. Quality is conformance to specification, while reliability is quality over time. Reliability from a customer perspective is that "the product works the way it was supposed to work for its desired period of use." Before you start improving reliability you must be aware of what you are improving.

As a practical matter, most companies have quality improvement processes in place throughout their manufacturing process. Understanding the difference between quality and reliability is the key because if you don't understand the difference you may just be improving your product quality, while not impacting your product's overall reliability at all. How can you separate the two?

First, focus on what's been going wrong in the manufacturing processes. Then take a look at the data that describes what has been going wrong. Review the data to see how the problem has been corrected. If it has been corrected by anything other than a design or process change, all you did was manage the process to maintain conformance to a specified parameter. You maintained conformance to specifications. If the improvement was made by an added inspection step, measurement step or any other added human intervention, this is a clue that all you did was improve the quality. If, on the other hand, a change was made, which improved the ability to maintain conformance to specification, in all likelihood you made a reliability improvement. If that is actually what occurred, current data will show that there was little or no

Improving Product Reliability: Strategies and Implementation. Mark A. Levin and Ted T. Kalal
© 2003 John Wiley & Sons, Ltd ISBN: 0-470-85449-9

Figure 2.1 Overcoming reliability hurdles bring significant rewards. Courtesy of Teradyne, Inc.

subsequent corrective action required to maintain this quality specification. The data then, will show that the problem area has been greatly improved or completely corrected.

Study the data that you already have. Gather the data from the manufacturing process and from field failure data; this can have immediate impact on lowering warranty costs. Problems that can be corrected by design take longer. Very often the decision has to be made to apply your resources on new designs rather than on old or current designs. If it is believed that the current design will be in production for a long time, resources should be expended on making changes to this design. This is also the time to begin using in- and out-of-warranty repair data. Compare this data to the changes put in place that were intended to make these problems disappear. If some problems did disappear, because the data shows that they stopped reoccurring, then there probably has been a reliability improvement. If some problems did indeed disappear, while others crept into the data, that information indicates that the processes are not in control.

After gathering the data and analyzing the information of failures, corrections, and the related processes, there will be a better understanding of where to start the reliability improvement process. In most cases, it will be apparent that there are both design and process problems. Select from the data the low hanging fruit and take corrective action to make these easy problems disappear. These will usually return corrections that will yield quality and reliability improvements with small utilization of resources. The harder problems will take more resources and a longer time to rectify, but they oftentimes return greater improvements. The gains from improving product reliability is multifaceted, Figure 2.1.

2.2 INTERNAL BARRIERS

Installing reliability into a company can be a very difficult and trying task. *Expect to encounter barriers all along the way.* The barriers you will encounter will be both internal and external. The internal barriers will be the most difficult to overcome and they are real. The external barriers are less significant. We will show that many of them are based on perceptions that are invalid. We begin by first discussing the internal barriers and suggest ways to break them down. The internal barriers will seem insurmountable at the beginning. Many companies find that a couple of years after they implemented their reliability program, the improvements they have made in product reliability are not significant. Obviously, the smaller the organization the easier it will be to implement reliability into the organization. Perseverance is necessary. In time, you will be amazed at what happens within the organization once these barriers begin to break down. This phenomenon is similar to that of a long-term investment in a retirement account. In the beginning, the retirement investment is small and the amount routinely contributed does not appear to bring you any closer to that final goal. However, over time this amount becomes significant as it begins to grow exponentially. The same effect will be seen within the company once the organization begins to see the benefits of reliability.

The internal barriers are the most difficult to overcome in implementing an effective reliability program. We have found that those companies that are successful make it part of their core competencies. Selling reliable products will distinguish your company from that of your competitors. We begin by looking at these internal barriers. A summary of the most common internal barriers is listed below:

1. Resistance to change
2. Lack of knowledge about reliability in management
3. Lack of knowledge about reliability in engineering
4. Inadequate training
5. Management does not support the process
6. Capital resources aren't there to support the process
7. No adequate staff to work on the issues
8. Goals not well defined or set arbitrarily
9. Adequate time not put in the schedule for the process

10. Adequate time not put in the schedule to fix the problems found

11. Process for implementation not well defined

12. Engineers want to move on to the next design and not go back to make improvements and fixes on older products

13. The attitude that "It won't work for us".

2.3 IMPLEMENTING CHANGE AND CHANGE AGENTS

Not surprisingly, the greatest barrier to successful implementation will be the resistance to change. During the early stage of implementation, the resistance to change will be across all cross-functional organizations within the company. The greatest resistance to change will be experienced in the engineering organization. Engineers, in general, are very set in their ways. It won't take long before you start hearing these phrases from engineering "We have to do it this way to make it work," "The system wasn't designed to work that way," and "We have been designing quality products for years; how is this going to make things better?" One of the problems that we commonly see, again and again, is that engineers do not understand the difference between quality and reliability. Sure, you may be designing and manufacturing quality products that meet customer's expectations. This was the goal for most companies during the 1990s but more will be needed to stay competitive in the twenty-first century. Quality addresses the ability to meet the customer's expectations at the time of purchase. Reliability addresses the ability of a product to meet those expectations over time. If you are building quality products that do not meet the test of time, you have a reliability problem. This is the message that we need to drive into the organization; the way we are doing business today will no longer work in tomorrow's competitive environment.

A common and universal reason for the internal resistance toward implementing reliability is the lack of knowledge about the process for product reliability. Many employees prefer to work in their comfort zone. It is human nature to fear what you don't know. It is natural to oppose something where success can't be guaranteed. The only way to resolve the fear and discomfort about the reliability process is to remove doubts through education. As the organization becomes more knowledgeable about the reliability process, the resistance toward implementation slowly diminishes. There are two necessary requirements that are needed to deliver the knowledge of the process. The first requirement is to have a charismatic leader or champion, who is highly knowledgeable about the process and is a good communicator of what needs to be done. Salesmanship here can be of great value. The second requirement is that management needs to support the necessary changes required to implement the new process. Both these requirements are necessary ingredients for success. Most importantly, management support is an absolute necessity.

The champion should be the reliability manager who is responsible for implementing the reliability process. Most companies lack this individual. Companies must go outside the organization to find this person. Selecting the right champion will be the difference between success and failure. Not only will this individual need to be a good communicator and motivator but he/she must also blend well into the culture of the organization. One of first things the champion needs to establish is credibility in the organization. The resistance level will drop dramatically once the champion establishes credibility. One common mistake often made is to select someone inside the organization, who has been successful in a different area to be the champion for product reliability. Often, the quality manager is given this new responsibility. Unfortunately, most quality managers lack the required skills and experience needed to establish credibility. This is not to say that quality managers can't make excellent reliability managers. However, during the implementation phase when the new process is being developed, problems will arise that are best resolved by someone having experience in similar situations.

Selecting the right candidate is vital, but equally important is the way the individual is introduced to the organization. It is a guarantee that there will be resistance to the implementation of reliability; the way you introduce the reliability manager to the organization will either initiate the breaking down of these barriers or will increase resentment and resistance.

At a small job shop where I once worked, the boss called a meeting on the manufacturing floor and introduced me to everyone as the new reliability "consultant." All 50 employees listened as the boss made it clear what was about to happen, and the faster it happened the better. He went on to say that everyone's cooperation and teamwork with the consultant would speed his (my) departure. "In fact, your next pay raises are on hold and being paid to the consultant as of this day; so the sooner we get this 'reliability thing' going the sooner your next raises will be forthcoming." This was support from management all right, but this introduction made the hill I had to climb much steeper.

In a very large company where a high-cost item was only part of the whole corporation's output, the management was considerably more diplomatic. "Let's see how this reliability thing works and in four or five months we'll take a closer look at the progress." That's not management commitment.

The second necessary ingredient in the implementation process is commitment. The commitment needs to be in manpower, capital resources, schedule allotment, and in management. The management commitment must be at the highest level. In addition, the commitment should be part of the five-year planning since the first couple of years the payback may not seem apparent. Once senior management has made the decision to implement reliability into the organization, a meeting should be planned with middle management and outside consultants in which shared goals for implementation can be set and the foundation for implementation established. Some companies use weekend retreats for this meeting. The implementation of reliability should not be viewed as an experiment.

If the need to implement reliability is real, so too must be the commitment. Typically, when a reliability manager is hired, senior management is high on the possibilities and low on the belief that it will be successfully implemented. This disbelief is even greater with the support staff. This disbelief can be diminished if the method of implementation includes the best practices in reliability. Many individuals in the organization, both management and staff, will usually have a negative attitude toward the whole idea. While others in the organization may have prior experience with reliability implementation, there is a 50–50 chance that they too may have a negative attitude toward the whole process. Simply put, disbelief in success is high.

In general, the design engineers' feelings toward the process will be unified in a group consensus. The reliability manager will be looked upon as an outsider. The need for senior and middle management commitment in the process is, therefore, paramount to the success of implementation. On the surface, the staff will consider this to be the latest management fad that doesn't seem to last more than a year. The boss has done things like this in the past. They usually don't pan out so we'll just humor him and let this latest fad die on its own. This view can quickly be addressed by sharing with the organization the five-year implementation plan and the commitment in resources to achieve this goal.

2.4   BUILDING CREDIBILITY

The third necessary ingredient in the implementation process is establishing the internal knowledge of what product reliability is and how it can be achieved. Knowledge can best be achieved through routine training sessions. The training should proceed in a logical fashion. First, you need to establish a common understanding of why product reliability is of concern to the company. This is an excellent opportunity to discuss the missed opportunities (i.e., problems) in previous products and the products that were late to enter the market. Next, there should be training on the reliability process. What is different from the way you have developed products in the past? How does it impact different organizations? How will it benefit them, and what resources will be available to achieve the goal? It is recommended that training meetings or mini seminars take place on a routine basis.

The training should not be limited strictly to reliability. Other product development organizations like the mechanical group, circuit board designers, field service, marketing, component engineering, and so on should hold classes to communicate issues and guidelines that affect product reliability. One topic that could easily be started is a training session on things that have failed on the manufacturing floor and in the field. There will probably be a lot of data to help with the class preparation. The class could point out what the problems were and what was done to correct them; then to identify if the problems were design- or manufacturing-related. The staff could learn a lot from this session and this will help jump start the attitude change that this reliability effort is real.

The classes are especially important to new employees and will help build a stronger and more effective team. For example, you can select from training sessions on manufacturing guidelines, mechanical guidelines, maintainability and availability, serviceability, testability, thermal management, product life cycle, accelerated testing, Mean Time Between Failure (MTBF), Failure Modes and Affects Analysis (FMEA), Design Of Experiments (DOE), physics of failure, component reliability, mechanical reliability, and system reliability. Periodic training sessions (one every two weeks or so) will develop the required knowledge base for achieving product reliability. Use internal experts to teach the sessions. This training process communicates who are the resident experts in these particular areas. The staff will learn the names and faces of the experts and will seek their help when needed. Something serendipitous that the authors found was that these classes introduced personnel with cross-functional skills to one another. The meetings created an improved working relationship that didn't exist before.

2.5   PERCEIVED EXTERNAL BARRIERS

There are many perceived external barriers facing companies that want to implement reliability. These are the following:

1. Time to market
2. Product development cost
3. Competitors do not do it
4. No local test facilities
5. No local experts.

The first two barriers, the time-to-market and product development costs, are the primary arguments used by those opposing implementation of reliability. This is a very shortsighted view. This perception is quite the opposite of reality. As we discussed in Chapter 1, when reliability is implemented in a concurrent manner, total product development time and cost are reduced. The idea may seem apparent but the perception will be quite different early in the implementation process. One reason is that one output from reliability drives a need for design changes. These design changes are interpreted by the design team as being unnecessary and will cause further delays to the project. However, if the reliability activities are performed concurrently with the design process, then the design changes will be implemented to the program at the lowest cost. For example, doing an FMEA with a cross-functional, multidiscipline team prior to design layout will identify design issues that would not have been caught until the prototype build or some time after product release. The net result is a reduction in the number of revisions required prior to the engineering release. Likewise, performing a Highly Accelerated Life Test (HALT) prior to releasing the product to manufacturing will remove design problems before first customer ship. This, in turn, will greatly reduce your warranty costs and can help to reduce product recalls.

The third barrier, and another common misperception, is that your competitors are not implementing reliability processes as part of their product development. This perception becomes more and more inaccurate every day as more companies embrace product reliability. While you linger, your competition gains a competitive edge. Besides these advantages, faster product development times, lower product development cost, and lower warranty costs, all come with improved product reliability. Making reliability a strong component in your product has several very desirable results that actually dovetail into one another. You get greater brand equity through product reliability, which commands a higher premium for your product, which generates increased profit margins overall.

The companies that do have product reliability programs consider this as part of their core competency. From FMEA, HALT, and from previous programs, the lessons learned can be applied to future products. The lessons learned can be captured in a database and made available to everyone through a computer-based retrieval system. Because these databases can get quite large, there should be search engines capable of finding studies based on key words or subjects. Secondly, lessons learned should be summarized into a Design for Reliability guideline that is updated and communicated to the design community.

One method that companies use to learn of these best practices in product reliability (prior to implementing the reliability program) is to benchmark their competitors. While this is an excellent idea in concept, it may be quite difficult to implement in practicality. One reason discussed earlier is that companies don't discuss their reliability programs. Secondly, companies rarely publish information discussing how they achieve product reliability. Here, much insight can be gleaned from interviewing and hiring reliability expertise to help you get started. Some companies use outside consultants and test facilities for part of their reliability activities. These companies have strict nondisclosure agreements in place that prohibit outside sources from discussing these issues. Some companies may even consider buying the competitor's product and performing a tear down to learn how they achieve product reliability. Fortunately, or unfortunately, competitors cannot tear down your product to learn how you achieve product reliability – not yet anyway.

2.6   IT TAKES TIME TO GAIN ACCEPTANCE

After a short time, everyone could see that the reliability manager wasn't going to be deterred, and day by day some swung over and got on board. In a few months, they could see small successes and more people were persuaded to join the reliability improvement process. Shortly after that, there seemed to be an avalanche of enthusiasm for the new "reliability thing." The reliability manager was swamped with requests to implement his reliability processes on their specific assemblies. They saw the light and wanted to get the benefit. Some diehards remained skeptical and indeed were relative speed bumps to the growth of the process. These doubters have to be discovered and persuaded to help make the reliability process happen.

In every case, the new reliability process is on a critical path. Resistance to it is easy to mount, because early market entry is extremely important for profits. Adding any activity delays market entry. Sometimes, as an alternative, management decides to place the reliability process on a sidetrack so that the existing product development is not slowed down by anything, especially the untried reliability processes. Placing the reliability process in parallel to the regular product development flow can work, but the results will be very small and the savings in development time will not be realized. Later, the evaluation of the new reliability process will have earned little support, because the data to support continuing the reliability effort will be almost nonexistent.

Even when the reliability tools reveal designs that have to be corrected, the time and resources needed to implement the improvements cause delays that management may not want to tolerate. If the time to implement reliability fixes are not made part of the product development schedule, the reliability process is doomed from the start.

Engineers are intelligent people. They have a technical understanding of how things work. They often have explanations for how things work (or will not work) on the basis of an instant analysis of a situation. When presented with the concept of reliability engineering, their instant analysis often finds reasons why it just won't work in their environment. They believe they know all about how to design and mount an electrical component, cool a system or select a fastener, and so on. What most engineers have little knowledge of is the feedback from the field failure data that has resulted from their latest design. As a result, they believe that everything they ever designed is fantastic, when, in many cases, there are parts of their designs that could have been made more reliable if the reliability tools had been applied to their designs before final approval. They need to be approached slowly with the reliability tools of the trade. Today, many engineers move from project to project or even company to company and have little opportunity to learn of their design oversights. By the time the feedback is available from field failures, they have moved on to new things. It is understandable that educating the engineer is critical to the success of the reliability effort.

2.7 EXTERNAL BARRIER

Logistics can be a barrier to applying HALT and other reliability tests to the product development process. New product developers can rent these services and spend a week (typically) to uncover product weaknesses. There are many of these sophisticated test laboratories throughout the country and the world. Some have HALT capabilities that can help the new user of HALT to design stress test regimens. Qualmark is a HALT chamber manufacturer. They have several HALT facilities that are located in major cities that are available for product developers (Qualmark at www.qualmark.com). A list of companies that provide HALT services and HALT equipment can be found in Appendix A. By contacting HALT chamber manufacturers, you can find the most local test houses where you can perform HALT. This list grows and changes often. (We recommend that you use the list of HALT chamber manufacturers provided in the appendix.) Unfortunately, the test houses are few and scattered around the country. In most cases, the facilities will not be close by.

The cost and logistics issues to perform the HALT testing at an outside facility can mount up fast. Design engineers may have to travel far to get to the HALT facility often requiring air, hotel, car rental, and other travel expenses. This, added to the facility charges for HALT testing, can make the cost a significant barrier to improving product reliability. Distance to the test house is very important to consider (the closer the better). If you are fortunate to have a test facility nearby, then travel cost is not a significant factor. The HALT service providers are in high demand, so you will need to schedule and book the testing time needed – typically four to six weeks in advance.

An engineer's time, like everyone else's, is valuable. They do not like to spend a lot of time traveling to off-site facilities to do their work. Minimizing this factor can be a make or break part of the reliability test implementation planning.

When the HALT process is not well understood and HALT testing is outsourced to a distant test facility, there needs to be a knowledgeable person who ensures that all the preparation work is complete before traveling to the test facility. In some cases, the new product developers may have to hire a consultant or contract engineer who specializes in this testing to join the HALT process. These skills can either be bought or learned. Often, both are needed at the start. This adds to the cost and is an additional barrier to improved reliability.

There are consultants and HALT machine manufacturers who provide HALT training. These services can be made available either on-site or at training seminars. A list of consultants for HALT testing can be found in Appendix A.

If a company has determined that the costs for implementing these reliability tests are prohibitive, they should take another look at the possibilities. If the manufacturer's end product does not give the customer the reliability that is expected, then the costs of not doing HALT will be much higher. The cost may be a loss of the customer base that could eventually bankrupt the business. This could mean that looking at purchasing the HALT machine is a better, long-term solution.

The upfront HALT testing costs appear, at first, to be prohibitive. When compared to the impact on the business (increased warranty spending, product recalls, and lost customers), they are small. The returns from long-term reliability improvements that become part of the product reliability will more than offset the dollars spent on implementing HALT testing. The manufacturer will benefit by retaining customers and by developing new customers who insist on nothing but the best reliability.

3

Understanding Why Products Fail

3.1 WHY THINGS FAIL

When we talk about product reliability, we are describing the trouble-free time period before a product fails. A failure is anytime the product does not function to specification when the product or service is needed. There are degrees of failure, for example, a color television that only displays black and white images, or a remote control that can change channels by using the number keypad but not with the channel up and down control. The new shirt that quickly loses a button and hangs unworn in the closet constantly reminds the consumer of his dissatisfaction. These are not complete failures, but the effects they have on subsequent purchases are the same. An automobile that stops on the way to a job interview, a computer that crashes in the middle of tax preparation or a parachute that doesn't open are much more severe failures. These failures are often communicated to others and have a more devastating effect on profit and future market share. The list is endless; the degree of failure can be varied, but the negative effect on your business is the same. A dissatisfied consumer results in the loss of repeat business.

What causes these failures? They can be due to inadequate design, improper use, poor manufacturing, improper storage, inadequate protection during shipping, insufficient test coverage and poor maintenance, to name just a few. A product can be designed to fail, although unintentionally.

For example, a large and expensive industrial product requires a high amount of airflow to cool the machine. When the fan stops working, the machine fails. In this example, a 20-dollar fan caused a multimillion-dollar system to stop producing products. The failure results in your customer having a complete production shutdown with significant unrecoverable dollar losses to the business.

Design engineers should know this and expect it to happen because fan manufacturers specify fan-life expectancies. To achieve product reliability, we

Improving Product Reliability: Strategies and Implementation. Mark A. Levin and Ted T. Kalal
© 2003 John Wiley & Sons, Ltd ISBN: 0-470-85449-9

must ask the question "What will wearout before the end of the customer-expected product life and why?" By identifying the things that will fail in the field, design changes can be made to improve product performance or a maintenance program can be established.

Materials expand and contract with temperature variation. Larger temperature variations cause greater material expansion and contraction. The amount of material expansion and contraction can vary for different materials. The greater the amount of temperature variation or the bigger the difference in material expansion rates between materials, the greater the stresses will be. Stresses due to temperature variations can cause component or solder connection fractures that eventually lead to failure. Designers can mitigate these failures with better environmental control, improving attachment structures or by proper selection of mating materials themselves.

Manufactured assemblies are a commingling of many subparts. These parts are attached by a variety of fasteners. Often, the fastener is a screw and nut. In shipping or through normal use, the associated vibration(s) eventually accumulate, which can loosen the screw. Over time, failures in fasteners can cause larger failures. These breakdowns can be avoided by selecting a fastener that will not come apart in the expected environment. Perhaps the proper torque with a split ring lock washer would be a solution in some applications; sometimes a press fit pin will do the job. Through design changes, the poorly chosen fastener that slowly leads to an eventual failure can be removed from the possibility of causing a failure.

Materials are stored to failure. When we think of hard manufactured goods, we don't think of the parts getting stale while they wait to be placed in the manufacturing process. The meat and produce industry must move their products to the end customer rapidly; otherwise they will suffer losses through spoilage and pests. Other industries have similar concerns but usually to a lesser degree.

The grocery store places their products on shelves. When a new shipment arrives, the old product is rotated to the front of the shelf and the new product is placed at the rear. This is commonly known as rotating or facing the shelves. This ensures that some items do not rest on the shelf too long to spoil. This is done on dairy products everyday. The term used in industry is *FIFO, First In First Out*.

Many electronics components actually start to wear out right after they are produced. How soon after they arrive at the manufacturing location they are installed in the product and shipped to the customer can be important. These parts also have to be used on a FIFO basis to ensure that the decaying process does not accumulate to lower the part's life expectancy.

The tin-plated copper leads on electronic component parts will corrode if left on the shelf at room temperatures for several months. These corroded parts do not solder to the circuit board well and tend to exhibit solder-connection failures sooner than they would have if the corrosion was not allowed to occur.

Adhesives have short shelf lives. If not used for several months, many adhesives are susceptible to early failure. Sticky-backed labels are often purchased in large quantities to get good pricing. Often these labels are in storage for several years before the last ones are applied to the product. In the field, these old labels will usually fall off in a few months and as such their value is lost.

Products are transported to fail. Assemblies arrive at the customer's destination only to be found inoperative because something broke during shipment. This is often a function of shipping cartons or crates that were not designed to stand the stress of shipping shock and vibration. The shipping carton can be a major cost item in the whole cost picture of a product. Sometimes, manufacturers save on these costs only to suffer even greater losses from returned goods from the customer. Manufacturers should require that the shipping carton be part of the total design effort. This should include appropriate testing to ensure safe delivery.

Products can be tested, operated (and the list goes on) to failure. When a product has failed, the failure mechanism must be learned to determine the root cause of the failure. The design of the product or the process must be updated to remove the failure possibility from happening.

At a meeting in a very large medical diagnosis manufacturer's facility where there were 20 design engineers and managers assembled, a new chief of new product development was heading the meeting. The meeting was to begin the planning of a new model of a large medical diagnosis system that is found in nearly every hospital in the country, and in fact, the world. After some initial discussion, the new chief asked how much more reliable this system would be relative to last year's model.

There was some laughter; then a respected designer explained that the new system would, of course, be less reliable than the older model because there were many new features that raised the component count and therefore lowered the overall reliability. The room full of engineers agreed. To them, this was obvious. The chief remained quiet and scanned the men at the table. Then, he asked one question.

"How many of you here have a VCR to record television programs?" Everyone said that they did. Then the chief asked if they remembered the older, reel-to-reel tape recorders, even the simple audio ones. They all remembered them. He went on to discuss the problems that the older units had. Many chimed in with their own horror stories of when the tapes got wound up all over the place and one channel was out of commission and so on. Then the chief asked another question.

"How many of you have had problems with the newer cassette VCRs?" There was silence. It appeared that the room full of engineers and managers had experienced no failures in their units ever. This came as no surprise. The Japanese manufactured them. Reliability was expected. Then the chief made a startling remark.

"These new VCRs are more complex and do many more things. They record video, audio in stereo and hi-fi stereo, they edit, and so on. Then, he asked a question and the room fell quiet.

"So why do the new VCRs seem to run forever?" There was no answer coming from the table.

The chief claimed that the improved reliability of the current technology was due to sound design and the removal of faults that are inherent in the design itself. This can be accomplished by identifying and removing faulty manufacturing processes with rapid corrective actions on field failures. New tools need to be found that put sound designs into bulletproof manufacturing processes so that the customer sees no reliability problems.

That meeting was held in the early 1980s. Since then, the reliability of components (in general) has improved two to four orders of magnitude. Parts were often specified in failures per million hours of operation (the term used is Lambda). Today, parts are specified in failures per billion hours of operation, which are referred to as *FITs* (Failures in Time). If parts were the main contributor to failures, then, with the vastly improved complexity of new devices, they would be failing constantly. We can all attest that they are not.

Televisions, radios, and automobiles all have more parts and last longer. This is due to the inherent design and the manufacturing processes, not the parts count. What is needed to improve the reliability of a manufactured assembly is to improve the design and the manufacturing process.

3.2   PARTS HAVE IMPROVED, EVERYONE CAN BUILD QUALITY PRODUCTS

Many things contribute to good quality and reliability. Nothing has been more important than the quality of the components that go into a product. Much work has been done over the past few decades to improve the quality and reliability of components. This effort has, for the most part, been very successful. In fact, the measurement used to describe the quality of components has been changed three orders of magnitude as well (from Lambda to FIT).

3.3   RELIABILITY – A TWENTY-FIRST CENTURY PARADIGM SHIFT

Today we hear from friends and colleagues and we know from personal experiences that the rules are different. With company takeovers, buyouts, mergers, and downsizing it is becoming clear that the old rules no longer hold up. The world is changing. Companies have to change to stay in business. Today's managers have to adapt their companies to these new paradigms.

What is a paradigm? The word has grown popular. It means a model or set of rules. If you do something a certain way, things will come out as expected.

Do this and you get that. Tip a glass of water and the contents pour out. Do it in space and it might not, since there is no gravity. The force on earth that makes the water flow down is not present in space, so the water does something new, something unexpected. In outer space, the rule that gravity pulls things down doesn't apply to the water in the glass. There is a new outcome when you tip the glass. Change the rule and you change the outcome. What we see in the marketplace is that the same old rules don't work any more. The paradigm has changed.

Paradigms don't change rapidly. Rules do; one at a time. Paradigms are what we believe to be true, not necessarily what really is true. Paradigms are made up of an assortment of rules. With more rules, the paradigm is more entrenched. With more established rules, our belief in the paradigm is stronger.

Five hundred years ago, the world was flat. It really wasn't flat but almost everyone believed and behaved as if it was. We think of the discovery that the world is round as an event in time, almost as if this transformation happened all at once. Then, after the new realization, we all behaved as if the world was always round. Not true. It took many years before the world was round (in everyone's mind). It took time before this new paradigm was well established. Now, we all agree that the world is round.

When many rules support the old paradigm, more obstacles have to be overcome to move into the new paradigm. More rules mean the paradigm changes slowly. At first, one rule doesn't work any more. It hardly gets noticed. Then a second rule changes, and another and so on. They begin to mount up. When many of the rules fall away, it becomes apparent that what used to work doesn't work any more. In space, what goes up goes up. For men in space, this is a new paradigm. So all the things that you did with water on earth have to be changed because the behavior of the world (in outer space) is different. The water will do its own thing even if you don't adapt. To make water work for you, you have to adapt. Your life will be better if you do.

Rules change on a continuous basis. Today, they change even faster. We must now learn new things faster just to keep from falling behind. When the rules of the market change, you had better take notice. The sooner you recognize that the world (the rules) is changing, the better your life will be. Why are so many people aware of paradigms now, weren't they there all the time? What has happened that has made paradigms become more important?

Information is abundant. There is television, CNN, magazines, and the Internet. Some are just databases or repositories of information. When combined for understanding, the result is knowledge.

If you know what your competitor is doing that allows him to sell at lower prices and still remain competitive, that's knowledge. How he manages to do this is better knowledge. Seeing what you are doing (or not doing) in comparison, can make or break your company. Adapting to meet the competition will keep you in the market place. Investing in new methods can take you past your competitor. This is what you need for your business to stay alive.

Do you ask yourself what is it that my competition is doing that allows him to sell below me and still maintain the margins needed to remain in business? It may be as simple as the fact that he keeps nearly all the money he receives from sales. He doesn't return a significant portion of the sales dollars back to his customers by way of warranty returns. He does this because his products deliver the promise of quality, day after day. His customers are satisfied and happy. His products are reliable. His customers reorder more of his product because they have experienced good quality for a long time. That's reliability.

His customer doesn't think much about what is helping him stay ahead of his competitors; he thinks about what are his major cost burdens. And then, he thinks of what to do about it to stay competitive. He works to improve or eliminate what it is that is hampering his business. And he reorders whatever it is that continues to deliver the quality and reliability he needs to stay in business. In effect, products that satisfy generate repeat sales with no added sales cost.

Another new paradigm is reliability. When your designs are mature and your processes are in control, the reliability of your product will be high. The return is in dollars not lost to warranty claims and upset customers. You, as a manager, have to make the changes that ensure quality and reliability. Otherwise the market will look to those who have learned these new rules earlier.

References

1. D. S. Steinberg, *Vibration Analysis for Electronic Equipment*, John Wiley and Sons, third edition, pp. 8–9, 2000.
2. W. S. Stewart, *Determining Bolt Tension, Machine Design Magazine*, November 1955.
3. R. W. Dicely and H. J. Long, *Torque Tension Charts for Selection and Application of Socket Head Screws, Machine Design Magazine*, September 5, 1957.

4

Alternative Approaches to Implementing Reliability

4.1 HIRING CONSULTANTS

The decision has been made to implement some reliability process on the new product. At the start, there is no one with skills in the company to begin making reliability improvements. Where to begin? Who to call? The first thing to do is to know what not to do. Do not locate a Highly Accelerated Life Test (HALT) facility; bring the product to be tested and expect optimum results. A little planning goes a long way.

Contact several contract-engineering agencies. They will often have resumes of engineers who have experience in HALT and other reliability processes. This will help surface local experts. Interview these specialists and determine if their experience is a good match with your product. During the interviewing process, you will learn of the local HALT facilities, if there are any. While looking for a HALT facility, you should inquire about: availability, flexibility, cost, staffing, and so on. Do these facilities have additional reliability-testing capabilities? Can they do Failure Analysis (FA)? FA will help you with in-process and field failures that have been difficult to correct.

When you contact these test houses, learn as much as you can about their capabilities and the costs of their services. They will usually have a fee for the test equipment, the test engineer, support materials, and so on. You will be surprised at how this fee will differ in one community. Often, if the test house is told that there is a considerable amount of business coming from more new products, then this long-term relationship might net lower pricing, even at the start. Visit the test houses. Meet the individuals who will interface with your personnel. Bring your engineers on these site visits. Essentially start a business relationship. You may have decided to hire one of the contract engineers you had interviewed; bring him or her too.

Improving Product Reliability: Strategies and Implementation.   Mark A. Levin and Ted T. Kalal
© 2003 John Wiley & Sons, Ltd   ISBN: 0-470-85449-9

4.2   OUTSOURCING RELIABILITY

Take the time needed to identify the scope of the long-term reliability goals and objectives. Then seek advice from consultants and experts at test houses. Tell the test house what you are planning to do. Get their input. They often can support and give guidance to your reliability improvement goals.

Seeking advice from consultants can be very beneficial, especially if the magnitude of the reliability improvement plan is large. This will require the guidance of a specialist. The specialist will usually be able to provide immediate training at your facility. This will also tend to speed up the start of the new reliability program.

This training should, at first, be at the top levels of the organization. This ensures understanding, commitment, and direction. Trained managers will then carry the message to the rest of the staff. This is important for continuity. Managers can select the reliability person(s), hire contract engineers and/or consultants, set the schedules, and track the performance of the reliability effort.

This is a general description of the early steps needed to get the reliability process started. There is more to getting reliability implemented than what is stated here, but as the reader continues, he will see that there are different strategies for different companies depending on their special needs.

Part II

Unraveling the Mystery

5

The Product Life Cycle

There are many choices that need to be made before a reliability program can be put in place. These decisions will have a significant impact on the organization and the time required to implement the reliability program. A small company will have different barriers and decision criteria than a large one. A large company that has made the commitment to improve its products through reliability may be willing to spend whatever it takes to be successful. Of course, this alone will not be a guarantee of success, and if implemented poorly, leads to lower profits with marginal returns on product reliability. A small company will have different constraints driving the need for more reliable products and will implement change in a different way from a large company. Before tailoring a reliability program that is best suited for you, an understanding of the reliability process, concepts and tools is needed. In Part II, we will unravel the mystery behind a successful reliability program.

5.1 SIX PHASES OF THE PRODUCT LIFE CYCLE

We begin this chapter with a brief overview of the reliability process. The process is the same for all companies implementing a reliability program. The degree to which this process is formalized will depend on the size of the company and the time-to-market constraints. (A more detailed description of the reliability process is presented in part four, "The Reliability Process for Product Development.") The reliability process needs to include the entire product life cycle. The product life cycle is a cradle-to-grave approach, where decisions made in any phase of the product life cycle will have an impact on product reliability, customer satisfaction, profit, and product image. In addition, the decisions made must consider the impact it will have on the life cycle of the product. The product life cycle consists of six phases. They are Product concept phase, Design concept phase, Product design phase, Validate

Improving Product Reliability: Strategies and Implementation. Mark A. Levin and Ted T. Kalal
© 2003 John Wiley & Sons, Ltd ISBN: 0-470-85449-9

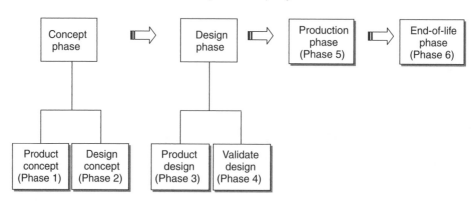

Figure 5.1   The six phases of the product life cycle

design phase, Production phase and End-of-life phase as shown graphically in Figure 5.1 below.

The reliability process is a multidiscipline effort that is conducted concurrently in each of the six phases of the product life cycle. The multidiscipline team consists of representatives who possess the knowledge, lessons learned, and expertise from their particular functional activity. Traditionally, concurrent engineering has been a multidiscipline approach that is implemented in the Design phase and Production phase of the product life cycle. Concurrent engineering, quality circles, continuous improvement, and other such programs have brought about a significant increase in the quality of products being manufactured in the United States. These programs are successful because they utilize the knowledge and expertise in the organization from all functional groups early in the product design phase, where decisions are made impacting product quality. As a result of these activities, manufacturers in the United States are producing some of the highest quality products in the world. These quality programs are more than 30 years old and are now practiced globally by competitors. Unfortunately, a strategy of competing on quality alone no longer provides a competitive advantage. Today's consumer is knowledgeable about product quality and can explain which products are of higher quality and why. These same consumers who have developed this understanding will soon be knowledgeable about product reliability. The time is not too far off when these same consumers will be capable of explaining which products are more reliable and why.

Easy and ready access to the Internet has helped further accelerate consumers' knowledge about product quality and reliability. Now, there are many search engines available where consumers can do comparison shopping for a particular product. These same search engines offer chat rooms where consumers can discuss questions and concerns that they have about a product. Some of these search engines have consumer product reviews in which you can read about a particular product of interest. Although it is not known to what extent the Internet and comparison shopping search engines have on the consumer decision-making process, they should be viewed as a powerful communicator of product quality and reliability. The Internet will provide a competitive advantage in the next decade for those who produce quality products that are reliable for the expected time of use.

The best way to design and manufacture quality products that continue to meet consumer expectations is to take a multidisciplinary, concurrent engineering approach to the product life cycle. It is also necessary to establish a reliability program that is integrated into the concept of the product life cycle. The reliability program considers any issues that relate to product quality and reliability along with any significant technology risk, which can impact a program. In the next section, we present an overview of the reliability program centered on the six phases of the product life cycle. Table 5.1 contains a summary list of the functional activities that take place in the product life cycle. Some concepts are introduced, which will be further explained in Chapter 7.

5.1.1   Mitigate Risk

Risk mitigation is a three-step process as shown in Figure 5.2. First, investigate to identify risk issues. Next, communicate the risk issues to all involved for acceptance of risk. Finally, develop a plan to mitigate the risk. Meet periodically to update status and review risk issues, close resolved issues and add new ones when they surface. Each of these three parts is described in further detail in Figure 5.2.

Investigate the risk

To identify the technology risk, each functional group reviews the product concept and identifies issues where the present technology, methods, processes, and tools will not ensure success. Early risk identification focuses the entire team on the concept of product reliability at the earliest. All risk issues, no matter how small need to be identified during this phase of the process. Low-risk issues will not have the same visibility and priority as high-risk issues. Therefore, to ensure that everything gets completed prior to first customer ship, all risk issues get captured at concept phase and recorded as shown below in Figure 5.3. Risk mitigation can be planned on the basis of each of these issues.

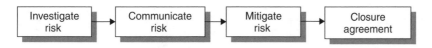

Figure 5.2   The ICM process

Table 5.1 Functional Activities for Cross-Functional Integration of Reliability

Functional activities	Concept phase		Design phase		Production phase		End-of-life phase
	Product concept	Design concept	Product design	Validate design	Production ramp	Volume production	
Marketing	Risk Mitigation – external VOC, reliability goals, lessons learned	Risk mitigation – internal VOC,	Implement risk mitigation plan				
Electrical design	Risk mitigation – external VOC, reliability goals	Risk mitigation – internal VOC, apply design guidelines	HALT, Implement risk mitigation plan, design FMEA	Risk mitigation closure, design & performance validation, operate FRACAS	ECO verification		
Mechanical design		Risk mitigation – internal VOC, apply design guidelines	Implement risk mitigation plan, design FMEA	Risk mitigation closure, design & performance validation, operate FRACAS	ECO verification		
Software design		Risk mitigation – internal VOC	Implement risk mitigation plan, design FMEA	Risk mitigation closure, design & performance validation	ECO verification		
PCB design		Risk mitigation – internal VOC, apply design guidelines	Implement risk mitigation plan, design FMEA	Risk mitigation closure	ECO verification		

Reliability	Risk mitigation – external VOC, technology risk, reliability goal, reliability process, lessons learned	Risk mitigation – internal VOC, lower level reliability goals, define reliability design guidelines, technology risk, reliability capital budget,	Implement risk mitigation plan, design FMEA, reliability estimates, root cause failure analysis, apply design guidelines, install FRACAS, HALT planning	HALT, risk mitigation closure, proof of design, proof of screen, root cause failure analysis, document findings into lessons learned, operate FRACAS	HASS, ECO verification, HASS, FRACAS, reliability growth	HASA, FRACAS, reliability growth	HASA, FRACAS
Software reliability		Risk mitigation – internal VOC	Implement risk mitigation plan, design FMEA	Risk mitigation closure			
Quality		Risk mitigation – internal VOC	Implement risk mitigation plan, design FMEA	HALT, risk mitigation closure, operate FRACAS	HASS, FRACAS, SPC, 6-sigma	HASA, FRACAS, SPC, 6-sigma	HASA, FRACAS, SPC, 6-sigma
Test engineering		Risk mitigation – internal VOC	Implement risk mitigation plan, design FMEA	Risk mitigation closure, operate FRACAS	FRACAS, SPC, 6-sigma	HASA, FRACAS, SPC, 6-sigma	HASA, FRACAS, SPC, 6-sigma
DFM		Risk mitigation – internal VOC	Implement risk mitigation plan, design FMEA, apply design guidelines	Risk mitigation closure, operate FRACAS, Process FMEA	FRACAS, SPC, 6-sigma	FRACAS, SPC, 6-sigma	FRACAS, SPC, 6-sigma

(continued overleaf)

Table 5.1   (*continued*)

Functional activities	Concept phase		Design phase		Production phase		
	Product concept	Design concept	Product design	Validate design	Production ramp	Volume production	End-of-life phase
DFT		Risk mitigation – internal VOC	Implement risk mitigation plan, design FMEA, apply design guidelines	Risk mitigation closure, process FMEA, operate FRACAS	FRACAS, SPC, 6-sigma	FRACAS, SPC, 6-sigma	FRACAS, SPC, 6-sigma
Manufacturing		Risk mitigation – internal VOC	Risk mitigation, design FMEA	Risk mitigation closure, process FMEA, operate FRACAS	FRACAS, SPC, 6-sigma	HASA, FRACAS, SPC, 6-sigma	HASA, FRACAS, SPC, 6-sigma
Customer support		Risk mitigation – internal VOC	Implement risk mitigation plan, design FMEA	Risk mitigation closure	FRACAS	FRACAS	FRACAS
Material management		Risk mitigation – internal VOC	Implement risk mitigation plan, design FMEA	Risk mitigation closure	FRACAS	FRACAS	FRACAS
Component engineering		Risk mitigation	Implement risk mitigation plan, design FMEA	Risk mitigation closure	FRACAS	FRACAS	FRACAS
Safety & regulation	Risk mitigation – external VOC	Risk mitigation	Implement risk mitigation plan, design FMEA	Risk mitigation closure	FRACAS	FRACAS	FRACAS

Note: DFM: Design For Manufacturing; DFT: Design For Test; ECO: Engineering Change Order; FMEA: Failure Modes and Effects Analysis; FRACAS: Failure Reporting, Analysis and Corrective Action System; HALT: Highly Accelerated Life Test; HASA: Highly Accelerated Stress Audit; HASS: Highly Accelerated Stress Screens; PCB: Printed Circuit Board; SPC: Statistical Process Control; VOC: Voice Of the Customer

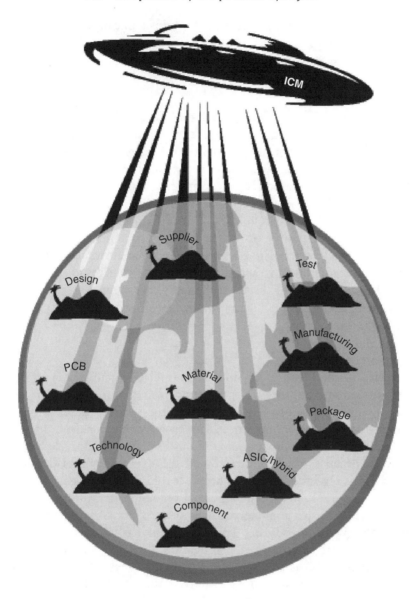

Figure 5.3 A risk mitigation program (ICM) needs to address risk issues in all aspects of the development program. Courtesy of Teradyne, Inc.

Risk identification is a process that must be documented and that addresses the following issues:

1. Description of the risk
2. Brief description of the activity needed to mitigate the risks
3. Impact of risk to program and other functional groups

4. Severity of the risk defined on a scale from insignificant to catastrophic

5. Alternate solutions.

Communicate the risk

Once the risk issues have been identified, they need to be communicated to the key shareholders of the program. They are the representatives of the functional groups that make up the concurrent engineering team. The shareholders should also include the senior management of the organization. Since the risk is shared, shareholders need to agree on each identified risk of the program and determine if it is necessary. The process of communicating the risk is best achieved in a formal meeting with all shareholders. The sole purpose of the meeting is to present the risk issues, agree on which risks are necessary and sign a formal agreement to resolve the key risk issues prior to first customer shipment. This agreement is a commitment to spend the necessary resources to resolve key risk issues and communicate to the team the risks that cannot be resolved in the scheduled time frame.

Mitigate the risk

The last step in the Identify, Communicate, and Mitigate (ICM) approach is the risk mitigation plan. The risk mitigation plan captures at a high level the activities that will take place to ensure that the product is reliable and can be manufactured at the volumes necessary to meet the marketing projections. The risk mitigation plan should not contain the steps necessary to achieve the deliverables, because it will be read by all the shareholders on the team and they may not be interested in all this detail. Included in the risk mitigation plan should be a list of all the experiments, environmental stress tests, tooling, and so on that will take place along with the desired outcome. The plan should also include a time frame during which each activity will start and close. The final necessary ingredient in the risk mitigation plan is the deliverable required for each activity. This is the one area in which many companies falter. The deliverable, as to what is required to have closure for each risk issue, needs to be clearly stated. For example, will a formal report, which contains all the necessary information for someone to repeat the activity and achieve the same desired outcome, be produced? Extremely high-risk issues will require an alternative path that is performed parallel to the primary effort but at a lower resource level. Finally, some of the risk activities will merge with other groups because the risk is common to both groups. In these situations, a team effort is necessary so that there is no duplication of effort and the results are agreed upon by the groups affected. The concurrent effort will ensure that there are no missed activities because it was mistakenly assumed that another group was doing that effort.

The ICM process requires an official sign-off, which allows the program to proceed to the next stage. This procedure is a gate that requires agreement that the risk is manageable in order to proceed to the next development stage. The sign-off can be either from the senior management team member or by the functional team members themselves. Often included in this sign-off is a definition of mandatory deliverables that must be resolved with an associated time frame usually aligned with a future phase of the product development process.

5.2   THE ICM PROCESS FOR A SMALL COMPANY

The reliability process for a small company may be different from that for a large or a medium-sized one. The small company often lacks the numerous functional support groups that are present in a medium and large-size company. The small company does not have the communication barriers that are present in their larger counterparts. As a result, a small-size company is often less process-restrained; it can solve problems informally and spend less time documenting and generating reports. The small company may face the same reliability issues and risks in developing technology-driven products that a medium- and large-size company may face. But because small companies are more flexible, they are able to respond fast to change and unfortunately take on added risk. The process for managing risk and ensuring reliability in product development is the same no matter how small the company is. Early in the product concept phase, an assessment needs to be performed to identify all significant risk(s). What is different in the small company is the way in which it mitigates risk due to its limited financial resources, technical expertise, number of qualified technical staff and lab capabilities.

In today's global environment, it is possible for a small company to compete with a large or a medium-sized company. The company that will win the market share will be the one to market first, with an effective business plan and the most reliable product. Since small companies still face many of the same risks and technology challenges as their large- to medium-sized counterparts, it is necessary to use the ICM approach to identify, communicate, and mitigate all significant risks. The small company needs to document the entire ICM process in order to ensure that no issues slip through without resolution.

The most significant difference for the small company is the implementation strategy for mitigating risk. Because the small company, in many cases, will be unable to adequately mitigate the risk internally, it should consider looking outside for assistance. Some of the alternative ways to mitigate the risk are as follows:

1. Hire a consultant with expertise in that area
2. Outsource parts of the work to a university
3. Hire temporary contract help

4. Technology licensing

5. Outsource to another company (not recommended that you outsource anything that is a core competency).

5.2.1 DFx – Design for Manufacturability (DFM), Design for Test (DFT), Design for Serviceability (DFS) and Maintainability, and Design for Reliability (DFR)

An important ingredient for successful implementation of reliability involves implementation of design guidelines. These guidelines include Design For Manufacturing (DFM), Design For Test (DFT), and Design For Reliability (DFR). These three guidelines, when implemented using concurrent engineering into product development, will ensure that the product will meet the minimum standard for manufacturing, test, and product reliability. An integral part of successful implementation of any "design for" guideline in product design and development, involves concurrent engineering. Concurrent engineering is a way of ensuring that things get done right the first time and that there is timely communication with all the groups involved in the design decisions. Concurrent engineering causes the developers of the product to consider all elements of the product life cycle, including manufacturability, serviceability, test, cost, schedule, user requirements, quality, and reliability.

5.2.2 Warranty

Every sales figure has buried in it a small dollar amount that is set aside to ensure that the seller has some capital available in the future to cover the costs of warranty claims made by their customers. The higher the sales the higher this dollar figure will be. More importantly, the lower the reliability, the greater the amount of revenue that needs to be set aside to honor warranty claims.

One way of looking at the contingency for the warranty claims figure is that this is the expected reduction in profit due to reliability escapes in the design and manufacturing process.

If the design calls for more nuts and bolts than needed, the cost of goods to manufacture the product will be higher. Cost-reduction efforts should be initiated to remove the unneeded fasteners to save some money on the cost of materials. After the unneeded screws have been removed, the cost of goods to manufacture the product will decrease and the profit will increase. The manufacturer then has the option of receiving a slight increase in profits or reducing the product cost with the opportunity of greater market share. Having options like these help ensure the company's future.

Warranty is like unnecessary screws. The more unreliable the product, the greater the amount of monetary reserves that have to be set aside to ensure that funds are available to cover warranty costs. It's like paying a little extra for the materials that went into the shipped product. Unfortunately, the manufacturer will pay in terms of warranty costs and in terms of lost sales never made because the customer takes his/her business elsewhere.

Companies with little or no reliability as part of the new product development product can have warranty expenses that reach 10 to 12% of the annual sales dollar (author's experience). Companies that have some reliability imparted during the development process can lower this figure to 6 to 8%. Only those companies that have implemented a cross-functional reliability process in their new product development process ever get this figure below 1%.

For companies with ten million dollars in annual sales and poor reliability, in essence, a million dollars is being handed over to the cost of doing business. To recoup this million dollars, how many salesmen would have to be hired to return this figure to the bottom line? How hard will the purchasing department have to negotiate to keep the cost of the product competitive? How many manufacturing people will have to be eliminated to maintain profitability? If this revenue was available for staffing more designers for product development, it would reduce time to market and increase the profit.

Every time you send a service person to your customer with replacement parts you are paying for poor reliability. All those extra parts in the stockroom that are there to support field service are really dollars set aside as a contingency for warranty claims, and poor reliability. All those materials parked in the stockroom are costing you money that can otherwise be actively making money, finding more customers, and hiring more employees.

Reliable products can help you prevent lost warranty dollars. To reverse this loss, the reliability of your products must be improved. And every improvement returns a portion of these lost warranty dollars. Not one, not two, but many reliability improvements will accumulate to return a significant portion of the lost sales dollar for better use.

References

Reliability process

1. G. Novacek, Designing for Reliability, Maintainability and Safety, *Circuit Cellar*, **January**(126) (2001) p. 28.
2. S. M. Nassar, R. Barnett, *Applications and Results of Reliability and Quality Programs*, 2000 Proceedings Annual Reliability and Maintainability Symposium, IEEE (2000).
3. R. Green, An Overview of the British Aerospace Airbus Ltd., *Reliability Process, Safety and Reliability Engineering*, British Aerospace Airbus Ltd., Savoy Place, London WC2R OBL, UK, IEEE (1999).
4. D. R. Hoffman, M. Roush, *Risk Mitigation of Reliability-Critical Items*, 1999 Proceedings Annual Reliability and Maintainability Symposium, pp. 283–287, IEEE (1999).

5. I. Knowles, *Reliability Prediction or Reliability Assessment*, IEEE (1999).
6. J. W. Evans, J. Y. Evanss, B. Kil Yu, Designing and Building-In Reliability in Advanced Microelectronic Assemblies and Structures, *IEEE Transactions on Components, Packaging, and Manufacturing Technology*, Part A, **20**(1), pp. 38–45 (1997) IEEE (1997) & Fifth IPFA '95 Singapore.
7. H. Caruso, *An Overview of Environmental Reliability Testing*, 1996 Proceedings Annual Reliability and Maintainability Symposium, pp. 102–109, IEEE (1996).
8. U. Daya Perara, *Reliability of Mobile Phones*, 1995 Proceedings Annual Reliability and Maintainability Symposium, pp. 33–38, IEEE (1995).
9. S. W. Foo, W. L. Lien, M. Xie, E. van Geest, *Reliability by Design a Tool to Reduce Time-To-Market*, Engineering Management Conference, pp. 251–256, IEEE (1995).
10. W. Gegen, *Design For Reliability – Methodology and Cost Benefits in Design and Manufacture, The Reliability of Transportation and Distribution Equipment*, pp. 29–31 (March, 1995).
11. W. A. Golomski, *Reliability & Quality in Design*, W. A. Golomski & Associates, Chicago, pp. 216–219, IEEE (1995).
12. D. J. Leech, Proof of Designed Reliability, *Engineering Management Journal*, pp. 169–174 (1995).
13. W. F. Ellis, H. L. Kalter, C. H. Stapper, *Design For Reliability, Testability and Manufacturability of Memory Chips*, 1993 Proceedings Annual Reliability and Maintainability Symposium, pp. 311–319, IEEE (1993).
14. J. Kitchin, *Design for Reliability in the Alpha 21164 Microprocessor*, Digital Equipment Corporation.

DFM:

1. D. Baumgartner, The Designer's View, *Printed Circuit Design* (January, 1997).
2. R. Prasad, Designing for High-Speed High-Yield SMT, *Surface Mount Technology* (January, 1994).
3. C. Parmer, S. Laney, DFM & T Guidelines for Complex PCBs, *Surface Mount Technology* (July, 1993).

6

Reliability Concepts

The information in this book is intended primarily for people in the design community, managers, CEOs, company presidents, associate reliability engineers, just about everyone except the reliability engineer. The reliability engineer understands the mathematics behind these reliability concepts that is of little importance to everyone else. Fortunately, the mathematics behind the reliability concepts is beyond the scope of this book. We, instead, will focus on the reliability concepts to provide understanding of what they are, how they get applied and how to interpret the results. You do not need to be a reliability engineer to understand and discuss these concepts. By understanding the concepts and tools, you will have a heightened awareness about product reliability and how it is achieved.

One of the more difficult challenges for someone wanting to implement a reliability program is developing an understanding of all the terms, definitions, and concepts used to describe product reliability. Many of these concepts are mathematically based and can be highly theoretical when pursued to minute detail. This has prevented all but a statistician or a reliability engineer from fully grasping these concepts. Obtaining this great depth of understanding is outside the focus of the book. However, it is vital to have a working understanding of the reliability terms, definitions, and concepts. One reason for developing such an understanding is that it will enable you to hire the right people when you begin to develop a reliability process. Once the reliability process has been established, these terms and concepts will be in common use when discussing product reliability. In some cases, we have oversimplified the explanation in order to avoid messy and confusing mathematics. In our opinion, it is important to understand these concepts because reliability is everyone's responsibility. As the organization becomes more knowledgeable about reliability, the products designed will be more robust and profitable to the company. Therefore, we present these concepts to develop this understanding without laboring over the

Improving Product Reliability: Strategies and Implementation.   Mark A. Levin and Ted T. Kalal
© 2003 John Wiley & Sons, Ltd   ISBN: 0-470-85449-9

mathematics behind them. Some fundamental reliability concepts commonly used in product development include the following:

1. The Bathtub Curve
2. Mean Time Between Failure (MTBF)
3. Warranty costs
4. Availability
5. Reliability growth
6. Design maturity testing.

6.1 THE BATHTUB CURVE

The most fundamental concept commonly discussed in product reliability is the "Bathtub Curve." The "Bathtub Curve," shown in Figure 6.1 is another way of looking at the cumulative number of failures for a product population operated over time. The bathtub curve is derived from the "Cumulative Failure Plot" shown in Figure 6.2. The cumulative failure plot is a plot of the running cumulative number of failures over time. For example, suppose you shipped 1,000 nonrepairable widgets and then kept track of the total number of widgets that failed through the life of a product. The plot will look similar to Figure 6.2. Often, in the first year, you will see a higher "rate of failure" for the product. Some of the more common causes are variations in the manufacturing process, using parts that have marginal tolerance, insufficient design margin or an inadequate test process. The failures that occur in this region are referred to as *infant mortality failures*. The failures due to "infant mortality" are considered quality related and are also called early life failures. The next part of the failure curve is called the *useful life*. The failure rate in the "useful life" region has stabilized and

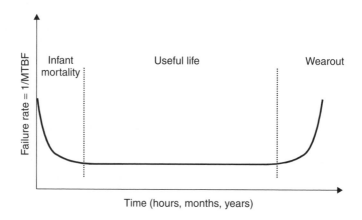

Figure 6.1 The Bathtub Curve (timescale is logarithmic)

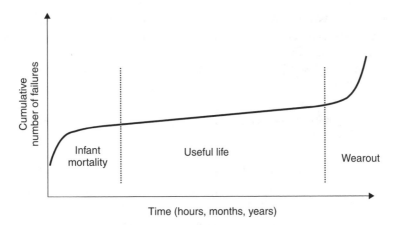

Figure 6.2   Cumulative failure curve

may be characterized by a relatively constant failure rate. The failure rate in this region is sometimes said to be *randomly occurring events*. After a long time, the product may exhibit a greatly increasing failure rate. This region of the bathtub is called *product wearout*. Here, the time has been reached when the product has consumed its useful life. It may be time for it to be replaced or upgraded.

The failure rate of a light bulb illustrates well the difference between useful life and wearout. Suppose a light bulb has a life expectancy of 2,000 h. The light bulb packaging shows this as the rated life (or median life). The 2,000 h represents the knee of the curve in wearout. The light bulb is not expected to operate after 2,000 h. However, incandescent light bulbs are extremely reliable; for this example, we will assume that it has a one million-hour MTBF. The MTBF represents the mean or average failure rate of the light bulb during its useful life. Figure 6.3 illustrates this difference.

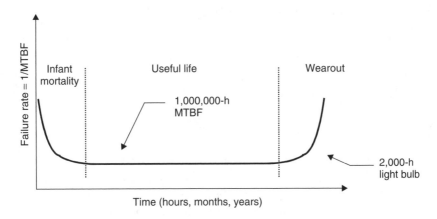

Figure 6.3   Light bulb theoretical example

Suppose now, you build a light panel to accommodate one million of these light bulbs described above. In addition, all the lights are wired together so that they operate at the same time, much the same way Christmas lights operate. During the useful life (the first 2,000 h of operation), one of the one million light bulbs will fail (stop illuminating) every hour. In reality, the light bulbs fail at a random rate. However, during the first 2,000 h, 2,000 of the one million light bulbs will have failed. This, by definition, is how the one million-hour MTBF is calculated. Now, if we continue to operate this light panel past the 2,000 h, the rate at which the light bulbs fail increases quickly. It may only take another 400 h before the majority of the million light bulbs no longer operate. This illustrates how the rate at which failures occur during the useful life is low, but once the product reaches its rated life, the rate at which the light bulbs begin to fail dramatically increases. In essence, failure is expected after 2,000 h, and so we can plan for this event. But, failures during the useful life are random and unexpected (one out of a million every hour).

6.2 MEAN TIME BETWEEN FAILURE

The most common term used to describe product reliability is the Mean Time Between Failure (MTBF). This term measures the failure rate of the product during its normal life. There are other ways to describe the failure rate of a product, which are explained in the following sections.

6.2.1 Mean Time Between Repair

Mean Time Between Repair (MTBR) is another way of describing the basic measure of reliability for a repairable system. It is a measure of the average time between all repairs for the systems in the field.

6.2.2 Mean Time Between Maintenances (MTBM)

Mean Time Between Maintenances (MTBM) is a commonly used term to describe the reliability of a repairable system. It is a measure of the average time between maintenance (preventive maintenance and repair) for all the systems in the field.

6.2.3 Mean Time To Failure (MTTF)

Mean Time To Failure (MTTF) is a commonly used term to describe the reliability of a nonrepairable system. MTTF describes the average time a collection of systems runs until the next system failure. This term is usually used in cases where the product will not be repaired. Because it is not repaired, it cannot have time "in between" failures in the normal operating sense.

6.2.4 Mean Time To Repair (MTTR)

Mean Time To Repair (MTTR) is the most commonly used term to describe the maintainability of a system. It is the sum of the time required to fix all failures divided by the total number of failures. The time required to fix the failure typically includes troubleshooting, fault isolation, repair, and any testing that is required to verify that the problem has been fixed. Simply stated, it is the time from when the customer could not use the product to the time the customer could use it.

6.2.5 Mean Time To Restore System (MTTRS)

Mean Time To Restore System (MTTRS) is similar to MTTR but includes the additional time associated with obtaining parts to fix the problem.

The above definitions are used to describe the frequency of events for a pre-defined environmental and operating condition. The frequency of these events can vary significantly for different environmental and operational conditions. There are many different ways to discuss a product's reliability. A company producing products globally will be more concerned about MTTRS because it includes the effectiveness of its field service and spare parts logistics. A company producing a disposable electronic product or a product with a short useful life will measure its MTTF. The method you choose to measure your product's reliability may be determined on the basis of your customer's needs, the market forces, and your ability to collect valid data.

The Mean Time Between Failure (MTBF) is defined as the reciprocal of the failure rate in the "constant failure rate" portion of the bathtub curve. MTBF is usually described in terms of hours between failures. The MTBF does not include the infant mortality failure rate and product wearout. To illustrate this point, we consider the reliability of a group of printers. This group may have an MTBF of 10,000 h (this MTBF number was made up and does not represent the actual MTBF of a printer). This statement implies that the average failure rate of these printers is about 1 per 10,000 h of operation. However, some printers last much longer. There will be other customers whose experience is that their printers last much less than the average life. In fact, customer experience may vary such that the observed printer time to failure ranges between 6,000 and 30,000 h of use. The reason a printer that has a 10,000 h MTBF rating and actually operates for only a few hundred to a few thousand hours relates to the bathtub curve. The operational life includes the infant mortality, constant life, and the wearout phase of a printer. Along this time line eventually, all printers can be expected to fail. Wearout is not usually considered part of the useful life of the product even though it may be hard to know when the wearout phase began. MTBF should really address only the constant failure rate portion of the bathtub curve, so care must be taken when observing failures. Observed field

MTBFs can show a wide range of times-to-failure, while the average (MTBF) is typically determined from testing and field experience. Some printers last much longer then the average and some do not. MTBF is this runtime average that is often used to judge conformance.

6.3 WARRANTY COSTS

When manufacturers use the term MTBF with customers, they must be careful. Typically, the customer sees this number as a sort of guaranteed number. So, when a single purchased unit fails in less than the "specified MTBF time," some form of compensation is often sought from the manufacturer. In general, for a given product with a given MTBF, there will be some products that fail before the MTBF (average) failure rate and some that will operate beyond the expected MTBF. The occurrences of the failure events are random and cannot be predicted; they can be estimated. The ones that fail, at times less than the MTBF specification, usually impact warranty budgeting as well as customer opinion.

Manufacturers can (and should) use the MTBF number to budget their warranty costs. If a product has an MTBF of 10,000 h and the product is typically operated continuously, then a one-year warranty (8,760 h) may be appropriate. The manufacturer may absorb any warranty costs when there are early failures (i.e., under warranty). Knowing the MTBF will give the manufacturer a basis to budget the warranty reserve figure for a given product. Using this example, we will next look at a simple situation to get a better understanding of how MTBF impacts warranty costs.

First, MTBF is a failure rate average of many units in operation that were not all manufactured at the same time. It is the fleet average. Keeping it simple, we can look at 100 units that were manufactured in the same short time period manufacturing cycle, say one week. They were probably placed in operation by their many users at about the same time, so their individual accumulated runtimes will often be similar.

In this example, a group of units will exhibit a fleet MTBF that is close to the specified MTBF figure, say 10,000 h. So, this minifleet of 100 units will exhibit failures (randomly), where the average failure rate will be 10,000 h. Some customers will see no failures; while others will experience one or more failures. Without going into the detailed math (see Appendix B), there will be approximately 37 units that will not fail during the specified 10,000 h! These customers are the lucky ones selected by the randomness of failure events. Some users will be unlucky and may have two, three or even four failures in the same 10,000-h time period! A simple example follows.

Suppose you have a product with a 10,000 h MTBF and you just sold the first 100 units. During the first 10,000 h of use, the following can be estimated.

Table 6.1 Failures in the Warranty Period w/Different MTBFs (One Year has 8,760 h Total)

Warranty period	MTBF = 8,760 h	MTBF = 87,600 h	MTBF = 876,000 h
1 year	100 failures	10 failures	1 failure
2 years	200 failures	20 failures	2 failures
5 years	500 failures	50 failures	5 failures

1. A failure about every 100 h. This is the average (mean).

2. A total of 100 failures in the 10,000-h time period.

3. The occurrence of failures is randomly distributed.

4. Some units (26) will have more than one failure.

5. Some units (37) will exhibit no failures.

The mathematics behind these estimations is presented in Appendix B.

How can you avoid much of this grief? Use reliability techniques so that your MTBF is very high. Even a product with a one million-hour MTBF may exhibit a first failure (in a fleet of 100) in about 10,000 h. But there will be 99 very happy customers because the next failure isn't (statistically) expected for another 10,000 cumulative hours.

A warranty requires a set-aside of sales dollars, intended to absorb warranty costs, during the warranty portion of the sold product. A product unit will experience one failure, on average, for every accumulated MTBF time period. Manufacturers must take this expected failure accumulation into consideration so that there are funds available to absorb this expected cost.

Manufacturers who sell products that have very high reliability and therefore high MTBF figures typically have very low warranty costs. They have happier customers who reorder products from this same manufacturer time and again. When we consider the number of failures in this example, all the failed units will be quickly repaired and placed back into service. Table 6.1 illustrates how MTBF relates to in-warranty failures for several warranty periods for repairable units.

See Appendix B for a more detailed discussion on MTBF and warranty budgeting.

6.4 AVAILABILITY

When a system fails and becomes dysfunctional to the customer, the time it takes to rectify the problem is critical (MTTR). Long repair times take a lot of productivity out of the usefulness of a continuously operating product, so having short MTTRs is a critical availability issue. Simply stated, static availability may be expressed as

$$\text{Availability} = \text{MTBF}/(\text{MTBF} + \text{MTTR})$$

Example: With an MTBF of 10,000 h and a 10-h MTTR:

$$\text{Availability} = 10,000/(10,000 + 10) = 10,000/10,010 = 0.999 \text{ or } 99.9\%$$

The reader can see that bigger MTBFs can drive availability to nearly 100%. Small MTBFs drive availability down, even with the same MTTR figure. A customer who experiences frequent outages needs to correct the problem quickly. If the MTTR is very short, less than 1 h, the impact is relatively small. But when the MTTR is long, such as several days, the availability can become very small. This customer will be unhappy. It is in the manufacturer's best interest to have high MTBFs, but they must also drive toward short MTTRs. Because parts availability is a major contributor to bringing a product back into operation, addressing this issue early in the reliability improvement planning is recommended.

Short availability is not a savior when the MTBF is low. Some manufacturers produce complex equipment that requires skilled, on-site, service personnel. If they can fix everything in a few hours on average, but there are many units operating, they will be fighting an uphill battle. For illustration, we will consider a customer who has 100 units in operation. With an MTBF of 1,000 h and a 4-h MTTR, there will be a new failure every 10 h on average. Service will have to fix these failures that are expected to accumulate at about 2.4 failures per day if the equipment runs 24/7 h. The repair person can be expected to fix an average of 2 units per 8-h shift and have another 0.4 units to repair before going home. The repair service may require more service personnel. Logistical constraints often drive MTTR requirements even higher. Long procurement from a few hours to a few days or even weeks may be observed. With an average of 2.4 accumulating failures per day, it is easy to see that the repair team will be quickly overwhelmed.

As failures appear, customers may have reduced output from their production. The customers will consider other manufacturers the next time they select equipment for the production line.

Shorter MTTRs translate into service capabilities that can correct field failures quickly. This means trained service personnel and readily available spare parts. Service/repair personnel have to be available to the customer almost immediately. This can be accomplished in several ways:

1. On-site manufacturer service personnel
2. Customer trained service personnel

3. Manufacturer training for customer service personnel

4. Easy-to-use service manuals

5. Rapid diagnosis capability

6. Repair and spare parts availability

7. Rapid response to customer requests for service

8. Failure data tracking.

6.4.1   On-site Manufacturer Service Personnel

Customers, particularly new customers, often do not have employees who are knowledgeable in troubleshooting, diagnosis, corrective action, and spare parts selection for their newly acquired devices. Providing factory service personnel to the customer at the customer's facility enhances long MTTRs.

6.4.2   Customer Trained Service Personnel

Knowing that the customer will need trained personnel, the customer's service team training should be completed before delivery of the product.

6.4.3   Manufacturer Training for Customer Service Personnel

The manufacturer should provide training to the customer's service personnel. If the customer prefers "before delivery training," it can be done at the manufacturer's facilities. To save the customer money, some training might be provided at the customer's facilities with prototype systems. Customers often desire this added service but it may be at the expense of not having a physical unit to work on, if their own unit has not yet been delivered.

6.4.4   Easy-to-Use Service Manuals

Part of the new product development process for increased reliability is the Failure Modes and Effects Analysis (FMEA) process. One of the deliverables (outputs) of an FMEA is a detailed understanding of how the product can fail. This information is invaluable to the service manual writers. They can use the FMEA to create a service manual that considers all the issues discussed by the design team. Service manuals will vary greatly depending on the end product but the FMEA process will always be an abundant source of information to help in the development of the service manual.

Manuals should be readily available to the customer. Many companies now place the manuals on the Internet. Providing a place on your web page to link to these manuals will shorten the MTTR of the end product. Placing manuals on CDs is another plus.

6.4.5   Rapid Diagnosis Capability

Factory training, along with the service manuals, can be very useful but some special designs can be difficult to diagnose. This may mean that special tools, test equipment, and software are required to help speed the diagnosis. Part of the service manual should have a recommended list of tools, and so on that should be available for rapid diagnosis. Expensive tools or special tools that can be obtained only from the manufacturer often hinder, rather than help in driving down the MTTR. Even a common tool that can be obtained almost anywhere in the United States may not be very easy to find in other parts of the world.

6.4.6   Repair and Spare Parts Availability

Wear items may need to be on hand at the customer's facility because of frequent failures. These include filters, fluids, lubricants, computer disks and tapes, and so on. A list should be available to the customer with a kit that should be available at the customer's facility so that the delays caused by these frequently used items are eliminated.

Other parts and subassemblies should always be in stock with the manufacturer for rapid delivery to the customer. A complete system of parts ordering, inventorying, and shipping should be in place to address the needs of the business. Globally dispersed customers require several parts locations to help speed their delivery to the point where needed.

6.4.7   Rapid Response to Customer Requests for Service

The parts and service department at the manufacturer's facility should be designed to make it easy for the customer to identify what is needed and the ability to obtain them rapidly. Special software and Internet capability may well be the solution. Here is where the quality of the naming and numbering of parts is very important. The customer may not have the same detailed knowledge as the manufacturer with the parts system(s) and may order the wrong part. This adds to the repair delay and greatly increases the MTTR. Make the parts ID system easy to use.

You must learn from the sales force if the customer feels that the system is useful. Find out what needs improvement and make changes accordingly.

6.4.8   Failure Data Tracking

Gather the data from the field to learn about troublesome areas. Use this information to eliminate problems. Often this information leads to design changes that can improve new products. These changes may save a lot of money and time during subsequent product development. Tracking the failure data

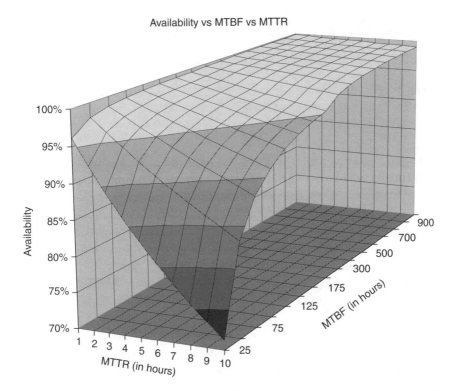

Figure 6.4 Availability as a function of MTBF and MTTR. Note: The curve has a slight ripple in it due to change in the MTBF axis. For the range between 0 and 200, it is marked in 25-h increments and in 100-h increments thereafter. This was done for resolution purposes to illustrate the impact of both low MTBFs and long MTTRs

everywhere can lead to rapid discovery and corrective action. A comprehensive Failure Reporting Analysis and Corrective Action System (FRACAS) will pay large dividends by lowering warranty costs and increasing the MTBF, thus lessening the impact of existing MTTRs.

The graph below (Figure 6.4) illustrates the relationship between, MTBF, MTTR and the resultant Availability.

6.5   RELIABILITY GROWTH

Reliability growth is the process of measuring the product reliability improvement when failure mechanisms are permanently removed. This was first described by J. T. Duane, who derived that as time passed the interval between failures increased. Simply stated, the MTBF increased after each failure mechanism was designed out of the product. For every defect that is removed from the system (meaning this failure cannot ever happen again in any system), the resultant MTBF must increase.

Table 6.2 Advantages of Proactive Reliability Growth

Proactive	Reactive
Rapid problem discovery	Data is more difficult to obtain
Failure data can be more accurate because the design team collects it	Data clouded by less-skilled service personnel
Equipment is easier to modify in-house	More costly to modify (recalls)
Can react to the first incidence	May need several occurrences
Manufacturer's reputation maintained	Manufacturer's reputation decreased

There are essentially two ways in which companies deal with reliability problems. The first choice is the one many manufacturers take, that is, to accumulate field failure information. When field data overwhelmingly indicates there is a problem, they investigate the failure to find the root cause. The problem then gets fixed and is closed with an engineering change. Hopefully, the design change becomes a recommendation for future designs. The total process from identifying a problem to implementing a fix may take considerable time, occasionally years. By then, the cost impact of the problem may have become significant. Because the resolution of the problem can take years, there is a good chance that new designs would be completed without the benefit of this knowledge. That is, the same old mistakes get designed into new products and cause similar problems. This is a form of reactive field failure analysis and corrective action. The process of finding failures and fixing them usually delays the product from reaching design maturity by a couple of years. Design maturity should ideally be achieved before the first units are shipped. Product recalls are the result of not identifying what will fail in the field before shipment. The second and preferred way to deal with reliability problems is to use reliability growth proactively as is shown in Table 6.2. The new product development process should have testing with data gathering (FRACAS system) that can help the designers find these failure mechanisms well before the design is finalized. Failure Modes and Effects Analysis (FMEAs) and Highly Accelerated Life Test (HALT) are two very useful tools for identifying potential reliability issues early in the design process. The reliability person should learn of and track every failure with the design person until the root cause is found and the design fix is implemented and validated.

6.6 RELIABILITY DEMONSTRATION TESTING

Reliability Demonstration Testing (RDT) is often confused with reliability growth. In the last section, we showed how reliability growth is used to bring a design to maturity before the first customer shipment. The more mature a design is at product launch, the lower the warranty cost will be. This is achieved by investigating all failures early in the design and keeping track of the company's corrective action to closure.

Reliability Demonstration Testing is a process that statistically ensures that the reliability goal set early in the program is met. There can be confusion regarding the order in which reliability growth and Reliability Demonstration Testing takes place. Reliability growth always precedes Reliability Demonstration Testing. We can start reliability growth measurements once the design is defined. FMEA and HALT identify problems; these surface and then get tracked using reliability growth techniques. RDT cannot start until the design is finalized and there is a final product to be tested. Put another way, reliability growth is a tool that drives up (improves) the MTBF of the product and the fruits of that effort are verified through RDT.

Earlier in this chapter, we showed that even though a system has a specified MTBF, the systems in the field may demonstrate a variety of different MTBFs. The MTBF we specify for the system is a measure of the "average MTBF" for all the systems. The actual systems in the field have failures at different times and at different rates. The MTBF number is the accumulated runtime for the entire population in the field, divided by the sum of all the field failures of the same population in the field. An individual product's MTBF varies because of the random nature of failure events. The random nature of these events can be mathematically modeled and may be used to construct the confidence interval. Think of the confidence interval as a part of a bell curve within which 90% of all the system's MTBF numbers will fall. Some will fall above and some will fall below this MTBF number. The further away from the specified MTBF number, the fewer will be the MTBFs.

RDT is the tracking of the accumulated product runtimes and the number of failures to verify that the product has achieved its MTBF goal. RDT is a way to show, through product testing, that the product indeed achieves the specified MTBF promised. RDT is a common method used to verify that the design meets a contractual reliability requirement. Reliability engineers are responsible for tracking the accumulated runtime hours against the number of accumulated failures. To verify that a system has achieved its stated MTBF, we will need to accumulate more product runtime without a failure. This total time is always more than the desired MTBF time.

There is a rule of thumb that can be applied to Reliability Demonstration Testing. The rule is used to verify that a system has achieved the specified MTBF goal at 90% confidence, that a system must run failure-free for 2.3 times the desired MTBF goal.

For example, we can say a system has a demonstrated MTBF of 1,000 h at 90% if it has functioned properly for 2,300 h without a failure. Recall that, when we talk about MTBF, there is some uncertainty specified around it. This rule assumes that the design maturity goal has been met with a 90% confidence

Table 6.3   RDT Multiplier for Failure-free Runtime

Confidence bounds (%)	Failure-free multiplier
95	3.0
90	2.3
85	1.9
80	1.6
75	1.4
70	1.2

level. If we wanted a higher confidence level, then the runtime without a failure will have to increase. Likewise, if the design maturity goal, can accept a lower confidence level, the failure free runtime will decrease. Refer to Table 6.3 to see the effect of the confidence interval on failure free runtime. A 90% confidence interval, typically, is widely accepted.

There is always a confidence level associated with Reliability Demonstration Testing. The confidence interval is the way to account for system variability and determine if it is acceptable. Let's imagine that a product has an MTBF of 8,760 h. With an exponential failure rate, there will be one failure every year. Of course, not every product will experience one failure every year. Unfortunately, life is not that simple. There will be some users who will have no failures in the first year time period and some who have one, two or more failures in the same first year. The variability is defined through a confidence interval. The confidence interval takes into account that not all users will have the same experience in the product's reliability. The way we deal with this random nature is through the use of confidence intervals to account for the variability in the reliability experience for RDT.

When a system or a number of systems in the field have demonstrated that their accumulated runtimes have reached at least 2.3 times the stated MTBF of 1,000 h without a failure, we can state statistically that there is a high degree of confidence (90%) that the product has demonstrated a 1,000-h MTBF. By using more than one system in the Reliability Demonstration Test RDT, the manufacturer can speed up the process. No manufacturer can state, however, *precisely* how long any individual unit will operate properly before failure.

What do you do when there is a failure during the design maturity testing to verify the product's MTBF goal? If this happens, does it mean that you cannot achieve your product's MTBF goals? No. There are still more statistics that can be helpful.

This point is illustrated in Figure 6.5 for demonstrating a 1,000-h product MTBF.

The x-axis is scaled for the number of failures. This test can be set up for one system or more. More is always better! The y-axis is in accumulated runtime hours. Again, this can be for one system or for more. When using more than

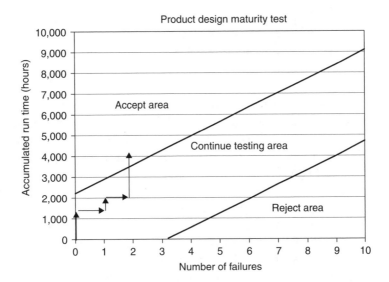

Figure 6.5   Design maturity testing – accept/reject criteria

one system to determine RDT, you must add the runtime hours and total the number of failures for all systems being used in the test. If you use two systems in the RDT, the accumulated runtime hours will collect at twice the rate as that of a single system. Using 5 or 10 systems will accelerate the RDT.

Figure 6.5 has some small arrows that depict what occurred when trying to demonstrate to perform a real RDT on a product that has a stated 1,000-h MTBF. The RDT begins at the origin, zero failures and zero accumulated runtime hours. For illustration, we show the first arrow as representing the test. This arrow lengthens and runs until there is a first failure at approximately 1,500 h. Then, the second arrow points to the first failure at the "number of failures" and extends until 1,500 h. The unit is repaired and the test continues and the third arrow climbs toward the accept line to represent this condition. When the unit reaches approximately 2,400 h, there is a second failure. Again, the unit is repaired and the test is resumed. Then, there are no more failures and the unit passes through the "accept" or "pass" line at approximately 3,600 h. This is the statistically required point in accumulated runtime hours that a system must run with two failures to have "demonstrated" that it has an MTBF of 1,000 h with a 90% confidence interval. If the system had no failures, the accumulated runtime hours needed would only have to have reached 2,300 h. As you can see, there may be times when doing an RDT that you might experience failures before the arrow passes through the "accept" line. This is still statistically correct. If the failures collect too rapidly and the tracking line passes through the reject line, then stop the RDT because the test has statistically proven with a 90% confidence interval that the reliability of the product is not capable of operating at the specified MTBF.

If the arrows never punch through the accept line, then you have more work to do in reliability growth before you should again try to demonstrate the final system MTBF. If the arrows soon punch through the fail line, then you must have decided to do the RDT prematurely or you simply have greatly overstated the system MTBF.

Can you place five units in the RDT for a while and then remove a few units and still have a statistically correct result? Yes; as long as the accumulated runtime hours and the total number of failures are recorded. Five units can speed the process. Removing a few units from the RDT will slow it down.

Can you place five units in the RDT and half way through remove some, install some new units and continue? Yes, statistically, but here it gets a little tricky. You must be certain that the infant mortalities have been removed from any of the new units. This is true for all the above examples. If the early failures are not removed, the RDT will, in all likelihood, fail. If you place units in and out too often, then the test may become invalid.

Can you perform a continuing RDT to ensure that the stated MTBF is remaining to specification? Yes. Place a group of units in RDT; remove them after a period of time for shipment; place new units in the same RDT; remove them for shipment; and so on. The idea is not to consume too much life from the units while testing them. The RDT test can be very challenging when the MTBFs are very long. For an MTBF of a million hours or more, you would have to demonstrate 2,300,000 h without a failure.

What is so magical about the number 2.3? It is simply the statistical correction number that acts as a multiplier for the 90% statistical confidence interval. If you wanted more confidence, the number would be somewhat higher and visa versa.

Reference

1. P. D. O'Connor, *Practical Reliability Engineering*, John Wiley & Sons, p. 360 (2002).

Reliability growth

1. J. Donovan, E. Murphy, *Improvements in Reliability-Growth Modeling*, 2001 Proceedings Annual Reliability and Maintainability Symposium, IEEE (2001).
2. L. Edward Demko, *On reliability Growth Testing*, 1995 Proceedings Annual Reliability and Maintainability Symposium, IEEE (1995).
3. H. Crow, P. H. Franklin, N. B. Robbins, *Principles of Successful Reliability Growth Applications*, 1994 Proceedings Annual Reliability and Maintainability Symposium, IEEE (1994).
4. G. J. Gibson, L. H. Crow, *Reliability Fix Effectiveness Factor Estimation*, 1989 Proceedings Annual Reliability and Maintainability Symposium, IEEE (1989).

5. J. C. Wronka, *Tracking of Reliability Growth in Early Development*, 1988 Proceedings Annual Reliability and Maintainability Symposium, IEEE (1988).
6. D. K. Smith, *Planning Large Systems Reliability Growth Tests*, 1984 Proceedings Annual Reliability and Maintainability Symposium, IEEE (1984).

Reliability demonstration

1. M.-W. Lu, R. J. Rudy, *Laboratory Reliability Demonstration Test Considerations*, IEEE (2001).
2. P. I. Hsich, J. Ling, *A Framework of Integrated Reliability Demonstration in System Development*, 1999 Proceedings Annual Reliability and Maintainability Symposium, IEEE (1999).
3. K. L. Wong, *Demonstrating Reliability and Reliability Growth with Environmental Stress Screening Data*, 1990 Proceedings Annual Reliability and Maintainability Symposium, IEEE (1990).

7

The Reliability Toolbox

7.1 THE FMEA PROCESS

The tool that is second only to Highly Accelerated Life Test (HALT) and Highly Accelerated Stress Screening (HASS) (these will be discussed later in the chapter) is Failure Modes and Effects Analysis (FMEA). FMEA is a very useful tool that can be applied without expensive equipment. In the late 1960s, the practice of using FMEA as a way to improve product design began to surface. It is a systemized series of activities intended to discover failures and recommend corrective actions for design improvements. These potential failures would otherwise not be discovered until the product was fully developed. The most important result of the process is that it will reveal a shortcoming before it is unintentionally designed into the product. In that respect, it is exactly like HALT and HASS in that it precipitates or identifies things that need changing in the design before the design is finalized. FMEA, like HALT and HASS, should be an integral part of the design process.

The FMEA process supports the design process by

- objectively evaluating the design through a knowledgeable team,
- improving the design before the first prototype is built,
- identifying specific failure modes and their causes,
- assigning risk-reducing actions that are tracked to closure.

In addition, the output of the FMEA can provide inputs to other key tasks. These include

- test and troubleshooting documentation,
- service manuals,
- Field Replacement Unit (FRU) identification.

Improving Product Reliability: Strategies and Implementation. Mark A. Levin and Ted T. Kalal
© 2003 John Wiley & Sons, Ltd   ISBN: 0-470-85449-9

Successful implementation of FMEA will

- improve the reliability and quality of products while identifying safety issues,
- increase customer satisfaction,
- reduce product development time,
- track corrective action documentation,
- improve product and company competitiveness,
- improve product image.

FMEA utilizes a team generally composed of the following sections:

- Design Engineering (mechanical, electrical, thermal, etc.)
- Manufacturing Engineering
- Test Engineering
- Materials Purchasing
- Field Service
- Quality and Reliability.

The FMEA is comprised of three sections: a Functional Block Diagram (FBD), a Fault Tree Analysis (FTA), and the Failure Modes and Effects Analysis (FMEA) spreadsheet.

7.1.1 The Functional Block Diagram

The functional block diagram is a step-by-step diagram that details the functionality of a development process. The process is broken down into three parts – input, process, and output (see Figure 7.1). The FBD is a high-level diagram detailing the high-level processes that take place for each input, process, and output. The steps identified under input, process, and output should not be highly detailed (see Figure 7.2). Each of the steps that are identified in the FBD becomes a process that is later evaluated using a fault tree analysis. For that reason, three to five steps are usually adequate to describe any input, process or output. Ten or more steps may be too detailed for the exercise and can bog down the subsequent FMEA.

Figure 7.1 Functional block diagram

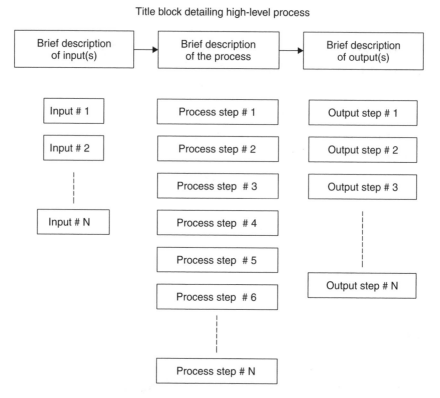

Figure 7.2   Filled out functional block diagram

The FBD details the outputs that are produced as a result of the processes that take place with the inputs. The output is the result of transforming the inputs via the process. Therefore, in Step 1, describe the process as a sequence of events. For example:

1. For a radio receiver, you turn the dial to a radio frequency and through an electronic process you hear the sound. (The electronic components comprise a series of circuits that, one by one, convert the signal at the antenna to an audible sound from the speaker. This series of signal conversions is the process in the FBD.)

2. A DC power supply has an AC power input, and through an electronic process it has a DC output voltage, for example, +12 V. (The AC from a wall outlet is converted into a varying DC; it is filtered, regulated, and sent out from the power supply as steady DC. This is the FBD of a simple power supply.)

3. An automobile transmission has a rotational force along an axis for the input, and through the drive shaft and mechanical differential it delivers this

rotational force to the drive wheels. (When the engine rotates, the force is transferred through a shaft and is coupled to the transmission through a clutch. The gears of the transmission select the amount of power to deliver to the drive wheels. Then, from the transmission drive shaft the power is delivered to a differential that finally connects the force to the wheels. This is an FBD of a transmission.)

A very basic FBD would be to describe the process needed to fill a glass tumbler with tap water. On the other hand, an extremely complicated FBD would be to convert nuclear energy into electricity using steam turbines. Even though these two examples are very far apart in complexity, they can be defined through an input, process, and the output needed to get the results.

The FBD can be as simple or as complicated as needed. However, the FBD should include all significant processes that are involved. A word of caution – it is not always desirable to define processes down to the component level. Keep the processes at a fairly high level; the simpler it is, the better in many situations.

Generating the functional block diagram

The FBD cannot begin until there is a technical understanding of the design or process by all the FMEA team members. Here is where the team leader can provide the necessary detailed information, that is, schematics, mechanical drawings, theory of operation, bill of material, and so on. For our example, we will use a simple flashlight and its schematic (Figure 7.3).

The schematic illustrates the components that make up a simple flashlight. There are two batteries, slide switch, bulb housing, bulb or lamp, housing spring, reflector, conductor from the lamp housing to the positive terminal of a battery, the spring for the batteries, and the flashlight housing (not shown).

The team must be sure that they agree that they understand the device described by the team leader and his documentation.

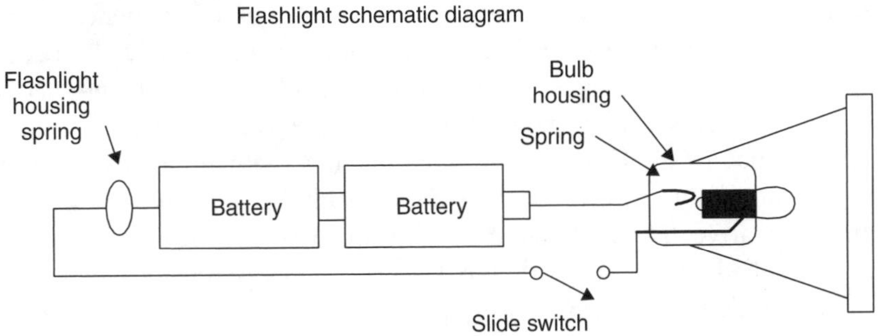

Flashlight schematic diagram

Figure 7.3   Schematic diagram of a flashlight

The FBD begins first by writing the three high-level labels: Input, Process, and Output. The labels can be scribed on Post-its®[1] and placed on the wall. The Post-its® are part of a toolbox that helps facilitate brainstorming exercises. Some additional items that should be part of the toolbox are the following:

1. Several packages of 3″ × 5″ Post-its®; (they are useful in brainstorming).
2. A large blank wall or large paper flip chart on an easel.
3. Several multicolored felt tipped markers.
4. A roll of masking tape.

The FMEA team leader first instructs the group to identify every significant process around which they wish to do an FMEA. This is a team effort and is done as a brainstorming exercise. The schematic flow diagram should be used as an aid. From the schematic flow diagram, identify each of the significant processes involved. (If a schematic flow diagram is unavailable, then the team will need to identify the major processes for the design through a brainstorming exercise.) Using Post-its®, have everyone create labels for all significant processes. Then, appropriately group the labels into common thoughts. Finally, review each of the grouped labels and agree that an FMEA should be done on that process.

There are several possible ways to approach filling in the FBD; we will look at two. Both approaches start the same way. First, identify the high-level processes that take place in the design. A schematic flowchart of the design can aid in identifying these high-level processes. Next, detail the process steps first; then identify the inputs and outputs required for the process to take place. The second approach works backwards, starting with the outputs for each high-level process. We will briefly describe the essential details of the two different approaches. They both work well and the appropriate choice is simply personal.

For the first approach, place each high-level process identified earlier as a title block for the functional block diagram. Then, write three FBD labels (Input, Process, and Output) on Post-its® and place the labels on the wall beneath each FBD title. Next, describe the processes that take place under the high-level title. For each high-level process, identify the process steps or sequence of events involved. Once the process steps have been identified, identify the necessary inputs for the high-level process to take place. The inputs are the ones necessary to support the process. Align them under the "Input" label. Finally, write down the outputs that result from the process, placing them under the "Outputs" label. An example of an FBD for the simple flashlight, is shown in Figure 7.4.

In the alternative approach, we start the same way by placing each high-level process identified earlier as a title block for the functional block diagram. We then write the three FBD labels (Input, Process, and Output) on Post-its® and

---

[1] Post-its® is a 3M registered trademark.

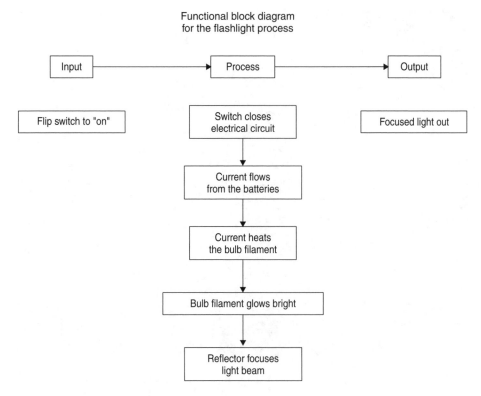

Figure 7.4 Functional block diagram of a flashlight

place the labels under the FBD title. In the alternative approach, we identify what the desired outputs are for the high-level process, write the desired output statement on yellow Post-its® and place it under the "Output" label. In the next step, we identify all necessary inputs required to achieve the desired output and align them just under the "Input" label. Finally, begin writing down the process steps needed to take the inputs and generate the desired outputs. Place these labels under the "Process" label.

The FBD is an interactive task that needs everyone's participation. Team members write labels on Post-its® and place them under the appropriate FBD blocks (Input, Process, and Output) as shown in Figure 7.5. The labels can then be moved around and rearranged with ease. As the activity progresses, you'll find team members rearranging many Post-its® until there is agreement on the FBD for each high-level process. After the FBD is completed, review each label with the team and make sure everyone understands the label and agrees with it. This step is often called *scrubbing the wall* and is intended to ensure that everyone understands and agrees with every item on the functional block diagram. When everyone is in agreement with the FBD, it is time to begin the fault tree analysis.

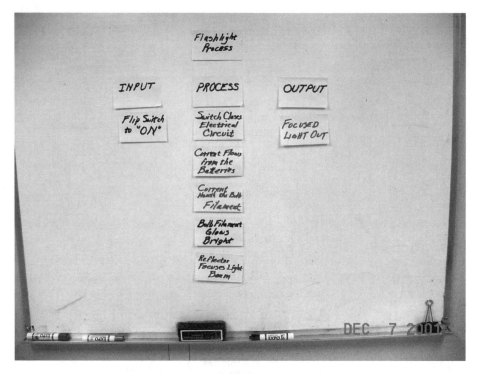

Figure 7.5   Functional block diagram of a flashlight using Post-its®

The FMEA process can consume significant engineering resources. One way to reduce this time is to prepare the FBD in advance. The top designer or team leader can prepare the FBD prior to the first team meeting. This will save time and speed the process up. It is best to circulate the FBD a week or two before the team's first meeting for team members to review. If the FBD is developed in advance, it should be reviewed at the first meeting to gain team agreement around which aspects of the design the FMEA will be performed.

7.1.2 The Fault Tree Analysis

The Fault Tree Analysis (FTA) is a logical, graphical diagram that describes failure modes and causes. The FTA diagram graphically shows all failures for a system, subsystem, assembly, Printed Circuit Board (PCB), or module. The FTA uses standard logic symbols (Figure 7.6), commonly found in flowcharting for process control, quality control, safety engineering, and so on, to tie together the sequence of events. The output from the FTA provides a better understanding of the causes that can lead to a failure mode. The results of the FTA can then be transferred to an FMEA spreadsheet. The FMEA spreadsheet uses the failure modes and their causes from the FTA and determines the effects of each

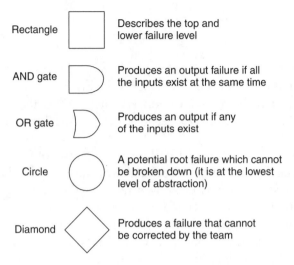

Figure 7.6 Fault tree logic symbols

failure cause on the design. The FMEA spreadsheet is also used to identify the most likely failure modes that will occur, as we will show later. The FMEA spreadsheet sets action plans in place to either reduce or eliminate the possibility of failure.

Building the fault tree

Begin the fault tree by stating the first failure mode from the functional block diagram. The failure mode is defined by taking the first item from the FBD (start with either inputs, processes, or outputs), and turning it into a failure statement. This is usually called the top event or high-level failure mode. If the FBD has as an output of "24 V provided to the output," then the failure statement would be: 24 V is not present at the output. Place the label "24 V is not present at the output" at the top of your fault tree. Then begin the brainstorming process to create a set of inputs that would be contributors to a failure that could cause the 24 V to not be present at the output.

Brainstorming

Brainstorming is a process in which everyone can contribute equally. The basic concept is that everyone sits quietly and writes down their ideas on yellow Post-its®. Obviously several members of the team will have similar ideas. Initially, this could be considered counterproductive or inefficient. However, the benefit that results from engaging everyone's participation is that more ideas will surface from the group. Set the ground rules for the brainstorming session as follows:

1. There are no bad ideas.
2. Everyone writes two to three labels and places them on the tree.

In brainstorming, it is agreed that there are no bad ideas. This way everyone feels comfortable in submitting his or her thoughts. Begin by having everyone write his or her two to three ideas on Post-its®. When the team is satisfied that everyone has recorded his or her ideas on Post-its®, it is time to start the FTA process.

Place the top-level (system-level) failure mode on top of the fault tree. Next, begin identifying failure causes associated with the above failure mode. You can go down several levels associating a second-, third-, and possibly fourth-level failure cause that is associated with the above failure mode. Place each subsequent failure cause beneath the previous failure cause using the Post-its® statements. To get to the next lower level of failure cause, ask the question, what event would have to occur to cause the higher-level failure? Usually, two to three levels of extraction in failure causes are adequate. The goal is not to drive to the root cause, but to bring to the surface failure causes in the design cycle that cannot be tolerated. Leave enough room between the levels for interconnecting lines and logic gates. Continue to do the process until the desired level of abstraction has been reached. See Figure 7.7 for a sample of a fault tree for the flashlight example.

The level of abstraction will significantly influence the amount of time needed to develop the fault tree analysis. If the team believes it is necessary to drill down to the most fundamental aspects of the design, then this is probably what

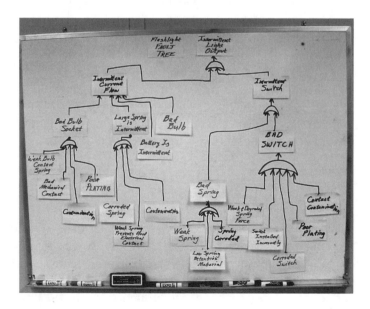

Figure 7.7   Fault tree diagram for flashlight using Post-its®

the team should do. However, the team leaders should use their expertise and knowledge of the design to prevent the group from going down an endless sequence of failure scenarios. All failure modes have causes. We circle the lowest level failure cause on the FTA and this will be the failure cause that is transferred to the FMEA spreadsheet (Figure 7.8).

It is possible, depending on the level of design complexity, for several failure modes to have the same cause. This is normal and is easily handled in the FTA exercise. Create several identical cause Post-its® and place them appropriately as inputs in the several failure mode statements. As you continue to build the FTA, it is easy to see why we recommend using a large wall to paste up the many Post-its®.

In building the FTA, you eventually reach a point where a decision needs to be made. The decision is that you have reached a sufficient level of failure cause description to evaluate its effect on the design. These failure causes are circled. However, you may reach a point where you cannot go further because the team lacks the expertise, knowledge or understanding of the failure cause. These failure causes receive a diamond because it will require outside expertise to resolve. All the lowest-level failure causes on the FTA should be either circles or diamonds. At this point, you are done with the fault tree analysis. Refer to Figure 7.9 for a simple failure mode and cause logic diagram.

7.1.3 Failure Modes and Effects Analysis Spreadsheet

The FMEA spreadsheet is a form that consolidates the FBD and the fault tree in a manner that facilitates organizing the relative importance or risks of the

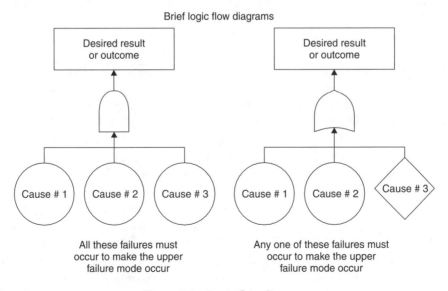

Figure 7.8 Logic flow diagram

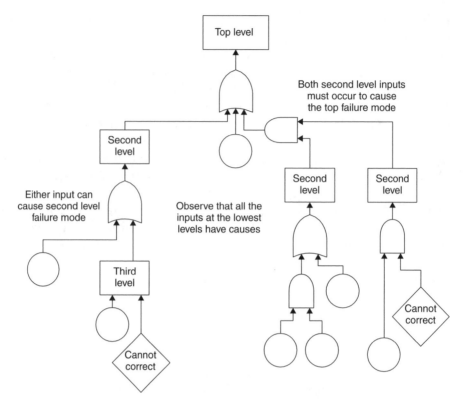

Figure 7.9 Fault tree logic diagram

failure modes. The FMEA Spreadsheet has several columns. The one used here has fifteen columns but the user can modify this form to suit the needs of the specific FMEA. The columns are described as follows:

A sample FMEA spreadsheet can be found in Table 7.1, which matches the above descriptions. The team leader should fill out the appropriate sections of the form. They are self-explanatory.

There could be more columns to fit the needs of the specific user of the FMEA tool. There are no hard and fast rules here.

The next stage of the FMEA process is to insert the failure modes and causes into the FMEA spreadsheet. This is actually the easy part of the FMEA process because the Post-its® rectangles from the fault tree are the failure modes. You merely paste them into the failure mode column. Insert the highest-level failure mode from the FBD in row one. Then, enter the causes from the fault tree under the next column. There may be several causes, so be sure to include them all for the specific failure mode. Remember that the rectangles from the fault tree are the failure modes and the causes are the circles and diamonds that feed into the rectangles. In the next column, enter the effects that were caused by the failure mode. Here too, there may be more than one effect. Do not fill

Table 7.1 The FMEA Spreadsheet

Failure Modes & Effects Analysis
FMEA#:   Company/Organization Name:
Assembly:
Owner:
Date:
Team Members:

#	Failure mode	Cause	Effects	Fault detection	S	O	D	RPN	H	FRU	Recommendations	Who?	When?	A

Legend: D: Detection Ranking; RPN: Risk Priority Number; S: Severity Ranking; O: Occurrence; H: Safety Hazards.

1. *Line or row number*: (We do not have this one and should add it to the form).
2. *Failure mode*: A brief description of the low-level failure mode.
3. *Cause*: What could cause failure to occur?
4. *Effects*: What effect does this failure have on the top-level design or process?
5. *Fault detection*: What could have been put in place to minimize or prevent the failure mode from occurring?
6. *Severity (S)*: A metric in units from 1 to 10, with 1 as minor and 10 as major. Severity is thought of from the point of view of the customer or end user.
7. *Occurrence (O)*: A metric in units from 1 to 10 with 10 the most frequent and 1 the least frequent. It is an estimate of the probable period before observing an occurrence; generally thought of as a field failure issue.
8. *Detection Ranking (D)*: A metric in units from 1 to 10, with 1 as a very high probability that the failure mode will be detected and 10 as a very high probability that it will not. (This can be confusing. The larger number represents a measure that is more difficult to detect.)
9. *Risk Priority Number (RPN)*: A metric that is the product of occurrence, severity, and detection ranking (just multiply the three together to get the RPN), this number can range between 1 and 1,000. The higher the RPN number the higher the risk of the failure mode.
10. *Hazard or Safety (H)*: Does this failure mode create a hazard? Does this failure mode create a safety problem?
11. *Field Replaceable Unit (FRU)*: Used to generate a recommendation for FRUs to your field service department.
12. *Recommended action (What)*: A brief description of what the FMEA team recommends will have to be done to mitigate the failure mode.
13. *Who*: The person or persons assigned to the recommended action.
14. *When*: The date on which the recommended action is to be completed.
15. *Audit (A)*: A check-off placeholder that indicates that the recommended action has been completed to the satisfaction of the FMEA team.

any more columns to the right; it is best to do that later. Take the next failure mode and enter it in the next line. Again, add causes and the effects. Continue until all the failure modes have been addressed. At this point, your rectangles, circles, and diamonds will have been completely consumed. See Table 7.1.

Ask what effect does this failure have on the rest of the system or process? In the flashlight example, an effect might be: no light output, or dim light, or the light gets weak very soon after turning the flashlight on, and so on. Complete the effects column fully. Then, move on to the fault detection column.

Now, beginning at the top, under the fault detection column, enter the mechanisms by which the failure modes could have been detected. An example might be: customer complains that the light gets dim too quickly, or life testing in a laboratory, validation testing or supplier qualification for a Design FMEA, and so on. Complete this column to the very bottom of the form.

Move on to the severity column. If the scale does not suit your specific need, then change it accordingly (refer to the RPN Ranking Table 7.2). If the severities are small, they are to be assigned small numbers. Severity rankings set impact. Severely impacted customer satisfaction gets larger numbers. Continue until the column is complete (refer to the RPN Ranking Table 7.2).

We will pause here to emphasize the importance of approaching the data entering one column at a time. As the FMEA team judges the levels for the various failure mode occurrences, if they stay in the occurrence frame of mind, their interpretation of what each number means will tend not to drift. If the team goes from occurrence, severity, and then to detectability, one parameter will tend to confuse the other. Because this is a very subjective measurement, it is best to avoid anything that may tend to impact the team's judgment.

Now in the Occurrence column, assign numbers between 1 and 10 that your judgment describes as the frequency of this specific failure mode (refer to the RPN Ranking Table 7.2). If the scale does not suit your specific need, then change it accordingly. Remember, it is the scale you'll use uniformly throughout this FMEA. Stay with this column until all the occurrence rankings have been entered.

Table 7.2 RPN Ranking Table

	Occurrence ranking		Severity ranking		Detectability ranking
1	Failure is unlikely or remote	1	Essentially no effect	1	Certain detection
2	Less than 1 per 100,000	2	Not noticeable by customer	2	Very probable detection
3	Less than 1 per 10,000	3	Noticed by discriminating customer	3	Probable detection
4	Less than 1 per 2,000	4	Noticed by typical customer	4	Moderate detection probability
5	Less than 1 per 500	5	Slight customer satisfaction	5	Likely detection
6	Less than 1 per 100	6	Some measurable deterioration	6	Low detection probability
7	Less than 1 per 20	7	Degraded performance	7	Very low detection likely
8	Less than 1 per 10	8	Loss of function	8	Remote detection likely
9	Less than 1 per 5	9	Main function loss, customer dissatisfaction	9	Very remote detection
10	Less than 1 per 2	10	Total system loss, customer very dissatisfied	10	Uncertainty of detection

Move through the Detection ranking columns in a similar manner. Things that can be easily detected get smaller numbers. If there is a very low probability of detection, the numbers will be higher.

When the severity, occurrence, and the detectability columns have been completed, the next step is to calculate the RPN by multiplying the three metrics together. The RPN number can range between 1 and 1,000. After you have completed entering all the RPN numbers, you will observe that the FMEA is beginning to take shape. Usually, there will be many numbers below a certain level or baseline, say 200. There will be a few numbers above that baseline as well. The magnitude of the RPN will highlight the top areas that need to be considered for improvement.

The next column, Hazard or Safety is used to consider if the failure mode could harm or cause injury to personnel. If the FMEA team considers this failure mode to be a safety issue, place a "Y" for yes in this column. Continue down the column with a "Y" or an "N" for no until the column is complete. Do not assign numbers here. This is not a metric that is to be multiplied together as part of the RPN. This is either a safety issue, or it is not and it should be treated appropriately. Some recommended action, to remove this as a safety issue, is needed.

In the next column, Field Replacement Unit (FRU), you can help determine if this failure mode can be best handled by fixing the problem in the field. If so, place a checkmark here under the FRU column. The FMEA process is not intended to be a tool that generates a list of FRUs. Here it is merely a tool that can be used to generate a recommendation for FRUs to your field service department.

The next column, Recommended Action, is where the FMEA can get bogged down. It is where the team makes a recommendation for change that will mitigate the failure mode. The team is not to determine a design change then and there. Each recommendation is to be assigned to a person who is an expert and can most efficiently deal with the failure mode. Usually, that person is a member of the team. Three columns can be addressed at the same time for they are the next improvement actions.

The "Recommended Action," "Who," and "When the task is to be completed" are to be discussed at the same time. Here is where the FMEA team "assignors" comes to agreement with the "Who," and sets a completion date agreed upon by all the FMEA team members. The FMEA team leader can manage the activities of each of those members who were assigned tasks in a normal fashion. The date is usually the date when the project needs completion of the recommended action.

The Audit column is for the reliability department so that they can track the reliability growth of the recommended actions from the FMEA. This ensures closure of all the recommended actions to the satisfaction of the FMEA team. Note in ISO 9001 companies and Biomedical companies, it is important to show a closed-loop corrective action system.

A sample FMEA spreadsheet can be found in Figure 7.11 that matches the above descriptions. The team leader should fill out the appropriate sections of the form. They are self-explanatory.

7.1.4   Preparing for the FMEA

To start an FMEA, the team needs a leader. This individual is most likely to be someone familiar with the FMEA process and can guide the team to success. It need not be a technical person. The first goal of the leader is to form a team that will collaborate to identify potential failure modes, their causes and correct the problems before the design is released. In preparation for the first meeting, the leader will assemble documentation that describes the design or process. The documentation is then distributed to each of the team members either before or at the first meeting. Often, it is desirable to submit this documentation a week or more in advance to give members a chance to familiarize themselves with the design. The following is a general list of the documentation needed in preparation for a Design FMEA:

- Mechanical drawings
- Electrical/electronic schematics
- Design process algorithms/software
- Process documentation that identifies inputs and outputs
- Miscellaneous items that describe the product and its function(s)
- An operational or functional block diagram.

The FMEA should be completed before the scheduled preproduction design release date. By the time the product is released for production, all the recommendations and actions assigned by the FMEA team should be completed, closed, and documented. Therefore, it is important to allow sufficient time in the design schedule for the FMEA process to be completed. Depending on the size of the assembly, the FMEA process typically takes (per assembly) 12 to 28 h to complete. It will take much longer to conduct FMEAs for large and complex products such as airplanes. The FMEA team should consist of cross-functional members who have been trained in the FMEA process. It is inadvisable to include team members who have not had FMEA training. Team members who are unfamiliar with the FMEA process will typically cause delays resulting from the fact that they do not understand the process. If there is an occasion where the whole team needs FMEA training, then a skilled coach who fully understands the FMEA process can complete this training task in approximately 4 h.

The first step of an FMEA is for the team leader to describe what the assembly is designed to accomplish. The simple way of doing this is to provide an FBD (functional block diagram) to the team. This diagram has basically three parts: inputs, outputs, and the functional process. The team is then tasked to review the FBD and to come to an agreement on how the assembly works.

The next step is for the team to identify failure modes and their causes. This is best achieved by using the brainstorming process. Any one of a number of process tools can be used to capture the team's findings. They are

- a fishbone or cause-and-effect chart,
- flowchart process,
- a fault tree analysis.

The fault tree analysis is the most common method used to identify failure modes. We, therefore, will describe the process for using a fault tree to identify failure modes and their causes. The design team may already be familiar with the FTA method as a way to determine possible root causes for a known failure.

The fault tree process begins with top-level failures, then second-level failures, third, and so on. The top-level failure represents the highest or most fundamental failure level. It can be as basic as turning on a light switch and the light does not go on. The second- and third-level failure represent events that take place as a result of attempting to turn on the light. Logic gates (i.e., AND, OR, NOR logic used in binary process flow) are then used to interconnect the lower level failure modes to the highest failure level, linking the entire process. At each level, the failure modes may have one or more causes. These causes can be mitigated by design change, manufacturing process steps, improved material selection, and so on. (In some cases, the causes cannot be addressed by the team and should be set aside for outside expertise.) The identified failure modes and their causes have an effect on the product. The failure mode is always something that will dissatisfy the customer to various degrees. These causes, when identified and corrected, greatly improve customer satisfaction.

Upon completion of the fault tree, a list of the failure modes and their causes has been constructed. At this point, more than half the FMEA process is complete. The next step is to fill out the FMEA spreadsheet. An example of the spreadsheet is shown in Figure 7.11.

In Figures 7.4 and 7.10, we built an FBD and FT for the Flashlight FMEA. Next, we will illustrate in Figure 7.11 how easily the Failure Modes and Causes can be taken from the FTA and placed onto the FMEA spreadsheet.

The determination of the RPN number comes as a result of the team assigning appropriate weightings for Severity, Occurrence, and Detectability. For brevity, the remainder of the FMEA spreadsheet is not illustrated.

When the team has completed the FMEA spreadsheet, the team leader either assigns actions to be addressed for each of the high RPN results and any Hazard or Safety related entries or determines that no action is required. The decision is based on the engineering resources available to fix problems and the severity

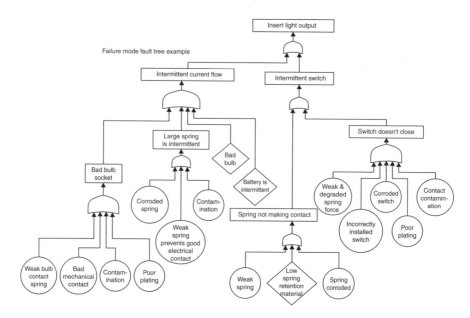

Figure 7.10   Flash light fault tree logic diagram

of the problems found. We have found that an 80/20 rule is a good guide in deciding what gets addressed. The 80/20 rule assumes that the top 20% of the RPN issues identified represent 80% of the potential problems. Keep in mind that all potential safety issues need to be addressed. All recommended action items should be completed, closed, and documented before the scheduled production release.

Earlier, it was noted that the FMEA process is second only to HALT and HASS. It is important to note that FMEA has several advantages over HALT and HASS. The HALT process is expensive. FMEA can be accomplished with very little expenditure other than the time used by the FMEA team. All the documentation used during the design review process can be used again in the FMEA.

The cost to introduce and perform the FMEA process is the same, regardless of the size of the company. The best part of an FMEA is that the cost to implement is small and independent of the size of the business. In addition, the resources expended on performing the FMEA will be recovered by reducing the total development time.

7.1.5   Barriers to the FMEA Process

Often, the greatest barrier to implementing the FMEA process will be getting the design community to accept the concept of reviewing someone's design for reliability in a systematic and detailed fashion. Most development engineers believe that they design highly reliable products. One reason for this is

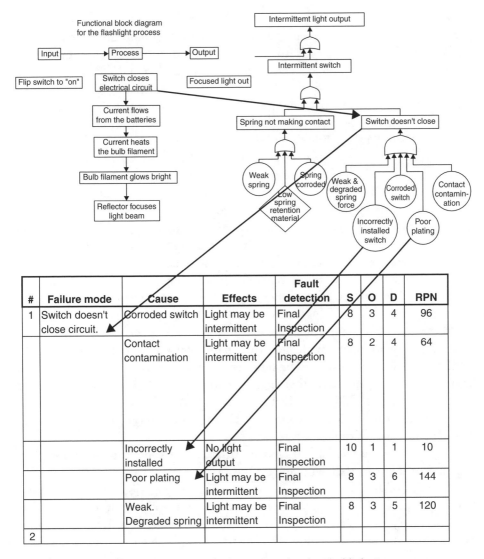

Figure 7.11 Functional Block Diagram for the Flashlight Process

confidence in their skills. Another reason is that most product development engineers are rarely aware of the field failures that resulted from their previous designs. Typically, design engineers go from one design task to another. Other engineers, often referred to as *sustaining engineers*, are responsible for resolving production and field problems. The disconnect between sustaining engineering and product development engineering is why mistakes from past designs are being repeated in new designs. If product development engineers are involved in resolving the original design problems, these design problems will not be repeated.

On one occasion, I pointed out to a design engineer that he was incorrectly mounting a component vertically (to save circuit board real estate space). This eventually leads to a failure caused either by shipping or by long-term vibration fatigue. The engineer replied that he had been designing that particular component the same way for the past 20 years without any problems. I asked the engineer if he had reviewed the failure reports that resulted from his past designs. He said he never had and that it wasn't his job. Then I asked him how often failures are reported. He said, "So far, never." This is a very large barrier called, *I've always done it this way*. This is a barrier that exists deep within most experienced design engineers. You'll have to overcome this barrier, one engineer at a time. The older, more experienced senior engineers will be the most difficult to change into rethinking the way we review designs for reliability. This does not imply that you should work on the younger engineers first. On the contrary, work on the senior engineers first. They will change and when they do, you will have allies that are already highly respected by the rest of the staff. It will make the changes easier to implement when you have their support.

Younger engineers and recent college graduates are much easier to persuade. They are not entrenched with their own particular set of tools and accept the FMEA process more readily. The older engineers tend to be more difficult, but their support will persuade other engineers to use the new process. There is one subtle advantage within the FMEA process that usually occurs around the first or second day of the FMEA process. Psychologists call it the "aha experience." Let me pause for a simple example.

Almost everyone has had to change a flat tire at some time in his/her life. When this happens, they will jack up the car and begin to loosen the lug nuts. When they do, the wheel will turn. Then, they have a fight on their hands. With one hand, they will hold the wheel to keep it from spinning in one direction, while they try to loosen each lug nut in the other direction. Eventually, they will get the lug nuts off, remove the flat tire, install the spare tire and refasten the lug nuts. The reason this usually happens is that we are rarely trained at replacing a flat tire. We are on our own the first time. Then one day, we have the "aha experience." When watching someone else loosening the lug nuts, we observe that this time the wheel is left low, touching the ground and then each lug nut is loosened just a little. This way they can use both hands on the lug wrench and let gravity hold the tire in place while it is still on the ground. Once all the lug nuts are loose, the car can be raised so that the tire no longer touches the ground. Now the lug nuts can be spun off with fingertip ease. "Aha." This will happen in the FMEA process, too.

Somewhere in the middle of the process, one engineer will observe a failure mode and its cause that he had never considered. It will come as a surprise. This is the anticipated "aha experience." It often comes from the most experienced engineer on the team. This golden moment is when that engineer is converted. This doesn't mean that he is completely on board, but his attitude has changed toward a more favorable direction in accepting the FMEA process.

Another barrier will be from management saying that the FMEA process will delay product introduction. This perception is not reality. If the FMEA is done early in the design cycle, it should not impact the design completion date. Some of the issues that will be uncovered in an FMEA would have surfaced during design verification, anyway, and led to project delays. In the past, when the FMEA process was not used, these issues hopefully surfaced in design verification. Then the engineering change process kicks in and the product is delayed while the fix gets implemented. The FMEA process can also save time because design engineers can spend a greater portion of their time on product development and less time fixing previous design problems. Design problems eventually get fixed. You can fix them before the first prototype is built when the cost is minimal or you can wait until your customer drives it.

Another significant barrier is that during the initial implementation phase, the FMEA process will take a long time to complete. This is normal. The process is complex and there can be a significant learning curve associated with implementation. At the end of the FMEA, have the team members note the strengths and weaknesses in the process so that improvements can be made. After you have implemented a couple FMEAs and implemented the process improvement suggestions, the process will be faster and proceed more smoothly.

Some feel that a good design review serves the purpose of finding all the design oversights and, as such, consider the FMEA to be redundant. Program managers are tasked with meeting delivery dates and will argue that the two processes are not necessary. Others consider the FMEA process as a replacement for the design review process. They argue that they should do one or the other but not both. Design reviews and FMEAs are two completely different processes with different goals and objectives. A design review is intended to ensure that the design requirements are met and that the documentation is complete and correct. The FMEA process is designed to discover failure modes and safety issues that cannot be allowed and to implement design changes to ensure that they will not surface. For reliability, you need to do both.

Where in the development process should the FMEA be done?

There is uncertainty as to where in the product development process the FMEA process belongs. Some development processes can actually support several FMEAs as the product is developed. This is especially true for large or complex systems. Each time a failure mode is discovered early, it can be more readily addressed at minimal cost. Waiting until the product is completely designed may not be the best policy. Doing one final FMEA will reveal all the failure modes at one time. The list may be too big to resolve due to pressures such as limited resources and time to market. On simple systems, the list of failure modes may well be more easily addressed. How many FMEAs and where they are performed will become clear as the user develops experience in the process and measures the returns for the effort.

7.1.6   FMEA Ground Rules

Keep the FMEA team engaged: Because the FMEA process usually takes several days or longer, many of the team members see a need to go back to their desk to tackle other tasks. The time set aside for the team should be long enough so that the FMEA can be completed, uninterrupted. If members leave from time to time, questions will arise during the process that only they can answer. Murphy's Law dictates that they will be away when they are needed to explain something. This delays the process. In the long run, the process will go faster if everyone commits to doing the FMEA without interruption.

Minimize interruptions: In some companies, where most of the engineers are very busy, and are often interrupted, it is best to perform the FMEA off-site, thus ensuring a minimization of interruptions. Even breaks and lunch periods can be optimally managed. There should be scheduled breaks. Break times should be short enough so that the members don't wander. Lunches should be catered so that everyone can begin again without the delays caused by stragglers.

Use the data available to assist in determining the level of importance. But don't stop if data is not available. Use the experience from the FMEA team. When discussions come down to opinions, seek outside information or help. Don't get caught up in endless opinion-dependent discussions.

Remember to do one column at a time until you reach Recommendations. This will help to avoid jumping to conclusions or making incorrect recommendations and it greatly speeds the process.

Maintain focus: It is easy to get distracted on a point and go down a path that doesn't add to the process. This is a time waster and the team leader should control this. Also, select a site free from outside interruptions.

Use the 80/20 rule: The resources needed to address each and every item may not be practicable.

Create a "Parking Lot" for issues: Marathon FMEA efforts are unproductive. The FMEA process can be mentally demanding, after 4 to 6 h, the team is likely to become mentally fatigued. It has worked well to limit FMEA sessions to 4-h time periods and then break. This allows the team to address daily business matters without having to distract the FMEA process.

The FMEA process is simple and straightforward. It begins with an FBD, then comes the FTA, and finally the FMEA itself.

7.2 THE HALT PROCESS

Without a doubt, the most important tool available to the product development and manufacturing process is a Highly Accelerated Life Test (HALT). There are many other methods that have been applied throughout the years to improve product reliability, but the HALT process has become the most effective and fastest method to improve product reliability.

HALT is an accelerated test designed to identify field failures before the first product is shipped. It is a method to apply stresses to a product while still in the design phase, which will reveal imperfections, design errors, and design marginality. After these design issues are identified, they can then be corrected through redesign. The HALT process is then repeated to verify that the design changes worked and that no new design issues resulted from the design change. The HALT process is very simple, yet few companies have fully implemented the process.

In fact, many companies do not have a reliability program in place. They consider their quality programs sufficient to achieve product reliability. These companies use the traditional approach of product development. That is, products are designed with "checks and balances" in place like design reviews. Design reviews check to verify the design is complete. The design review will verify, for example, that the parts list needed to build the design is complete. The design review may also verify that the material list is in a "standard format", usually defined by the manufacturing process. Design reviews typically verify design completeness through a concurrent activity involving all involved functional groups to review the documentation package and verify that the design is complete. Examples of some of the areas that are covered in a design review are as follows:

Engineering
- Schematics
- Block diagrams
- Theory of operation
- Outline drawings
- Input/output descriptions
- Thermal design
- Component derating
- Power descriptions.

Manufacturing

- Design For Manufacturing (DFM) guidelines
- PCB guidelines
- Material list [Bill Of Materials (BOM)]
- Assembly drawings
- Assembly instructions
- Manufacturing cost.

Test

- Design For Test (DFT) guidelines
- DFT cost
- Test software
- Fixtures.

Supplier

- Approved suppliers
- Material costs
- Delivery lead times
- Alternate sourcing.

Software

- Software debug and validation.

The list of items covered in a design review can be extensive. Most companies also use some form of continuous improvement to improve and streamline the design review process. The one component typically left out of design reviews is reliability, especially field reliability information. Design reviews may cover some reliability issues, but the issues are generally based on lessons learned. Issues such as derating, DFM and DFT improve product reliability and are sometimes covered in design reviews. However, reliability should be a bigger part of the design review process. The best way to reveal the reliability issues in a design is through HALT testing.

After the initial design is complete, a prototype is fabricated to test and verify that the design meets specification. Usually, not all the requirements are met and redesign is needed. Later, after the changes have been made, the redesigned prototype units are tested again to "prove out the design." The process of verifying that the design meets specification is referred to as a Design Verification Test (DVT). At this point, the design is considered complete and ready for production. If you perform HALT testing *before* the DVT, there is strong likelihood that you will pass the DVT the first time. HALT is not intended to replace DVT. By performing HALT on the first engineering units that are functional, reliability issues are identified and fixed early in the development cycle. The end result is a faster time-to-market and passing DVT testing the first time.

In the traditional approach, products are manufactured and shipped to the customer. In the first year or two of production, there is an accumulation of field failures that consume warranty dollars and often create dissatisfied customers. Teams are then formed that are dedicated to investigating the field failures, determining their root cause and developing corrective action. This is followed by an endless stream of corrective Engineering Change Orders (ECOs), to eliminate the problem. The design problems can take years to resolve and delay the product from reaching design maturity. This long delay will have its effect on profitability and customer satisfaction. The HALT process speeds up the product design cycle and significantly reduces the number of field failures typically experienced by early production.

The HALT process is an accelerated test, which will precipitate field failures in a relatively short time period, well before any product is in the field (refer Summary of HALT and HASS Results at an Accelerated Reliability Test Center, Mike Silverman, Qualmark Corp, Santa Clara, CA, 1998 Proceedings Annual Reliability and Maintainability Symposium). Once these failure modes are identified, they can be removed through redesign. Then, by applying the HALT test after redesign, the design fixes can be verified with the assurance that no new failure modes have been designed into the product. The end result is a final product that is free from defects while having a significant reduction in lost dollars to warranty claims. HALT is a stress process that accelerates failures so that they can be corrected before first shipment. HALT yields design maturity before the first unit is shipped. So HALT can be considered a "design maturity accelerator."

HALT testing requires that the device is powered up and operational while diagnostics monitor the device for normal operation. By monitoring the device under stress, failures can be detected along with the point-in-time and environment conditions when the device fails to meet specification. If a failure occurs, the Device Under Test (DUT) is removed from the HALT chamber and the next device is tested. Similar stresses are then applied to the next device to learn if it fails in a similar manner. While HALT testing is performed on the next device, the previously failed device is evaluated to determine the root cause of the failure. That device is then fixed and returned to the cycle for the next step in the HALT process. After five or six devices have been HALT stressed, a Pareto chart can be created of the failures. The Pareto chart (Figure 7.12) graphs five (hypothetical) precipitated failures and how many of each was discovered in the five test units.

In Figure 7.12, the design team might consider not dealing with failure E because it only happened on one unit out of five. This is often referred to as a

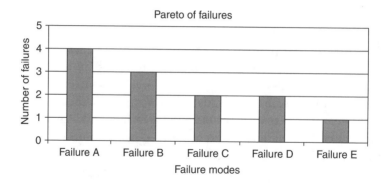

Figure 7.12 Pareto of failures

single event or anomaly and is of little importance. Not true. If 30 units had received HALT, there would probably be more failures in column A through E. There may well be even more failure modes. The point here is that with more units failure E would no longer appear as trivial; there could be many more assemblies with failure mode E. Because of the very small sample of five units in HALT, even the single failure mode E is significant. Failure analysis to the root cause is needed for every HALT failure because they are all likely field failure modes.

The full intent and purpose of the HALT process is to drive units to failure. Investigative techniques such as failure analysis, which drive down to the root cause of the failure, will reveal the true physics of failure. Once these failures are identified, they can be remedied by redesign.

Finding the root cause of the failure is critical. Just fixing the failure is of little value. A major automobile manufacturer discovered that some vehicles had completely dead batteries upon delivery to the dealer while others, on the same truck, were fine. Replacing the battery fixed the problem and allowed the dealer to sell the product. But what caused some of the batteries to fail? The solution turned out to be simple. Some cars, when loaded onto the delivery trailer, were at a significant angle. This caused the trunk lid sensor to activate the trunk light, but just for those cars at a steep incline at the rear of the delivery trailer. The trunk light went on because the sensor in the trunk lid turned the light on because the angle of sensor was correct for an open trunk lid. The solution was to disconnect the batteries before shipment to eliminate this problem. The failure mode was known and the root cause was discovered. The corrective action was acceptable for the short-term and a long-term solution was forthcoming. It was a lot easier to remove one battery cable and reattach it later than to replace batteries at the dealer.

Many years ago, radios and televisions had vacuum tubes. They were relatively unreliable in that their filaments would burn out, a major failure mode. The service person who replaced the tube fixed the problem but did not find out the root cause. Over the years, filament design extended the life of the tube technology by finding the root cause of filament failure and making changes that improved the product. Still, the tubes failed. Then, the transistor was invented and the failed filament root cause was solved. There are no filaments in transistors to fail.

7.2.1 Types of Stresses Applied in HALT

The HALT chamber is capable of applying two different types of stresses to the product, vibration and temperature. For the HALT test to be effective, these two stresses (at a minimum) are required. However, these are not the only stresses that can be applied to accelerate a product to failure. Examples of other stresses that can be applied in conjunction with temperature and vibration are as follows:

- Voltage margining
- Clock frequency
- AC supply margining (voltage and frequency)
- Power cycling
- Voltage sequencing.

The stresses described above can be applied individually at first and then in combination with the other stresses. The decision of which stresses to apply is based on experience and what is feasible. At a minimum, vibration and temperature are required. This point is illustrated in Figure 7.13. The graph is

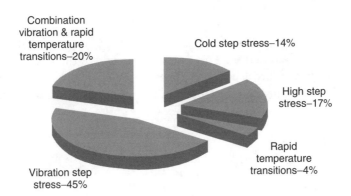

Figure 7.13   HALT failure percentage by stress type[2]

[2] HALT stress test failure breakdown. Mike Silverman "HALT and HASS Results at an Accelerated Reliability Test Center," IEEE Proceedings Annual Reliability and Maintainability Symposium, (© 1998 IEEE)

from the work done at QualMark Corporation, a HALT testing facility, and is a summary of their testing on 47 products from 33 companies and 19 different industries. The testing started with cold step stress and proceeded around in a clockwise direction ending with the combination of vibration and rapid temperature transitions. If only temperature testing was performed, 35% of the design failures would be identified. Likewise, if only vibration testing was done, 45% of the design failures would be identified. The power of combination stresses to identify design failures is evident. Temperature or vibration alone identify less than half of the reliability design issues. That is why it is important to apply both temperature and vibration to achieve the goal of accelerating the greatest number of field failures in a relatively short time.

Using accelerated stresses, first singularly and then in combination, will reveal reliability design issues that can be eliminated through redesign. The redesign is performed early in the design cycle where it is the least expensive to implement. After the design is corrected, it is necessary to retest the product with HALT. This will ensure that the fixes worked and that no new failure modes were designed into the product as a result of the redesign.

7.2.2   The Theory Behind the HALT Process

When a product is designed, it is tested to verify that it meets all its design specifications. Design specification may include: output performance, temperature, vibration, shock, power supply levels, duty cycle, frequency, distortion, power source limitations, altitude, humidity, temperature, and many more. We will illustrate this point with a single specification (i.e., temperature) and refer to it as having an upper limit and a lower limit (refer to Figure 7.14). This is the design operating range the product must meet in order to function normally and meet design specifications.

The design operating range, described in Figure 7.14 can be applied to all the design specifications. It is in the operating range where the end product is designed to properly function. This is the range where the development team tests the product to ensure acceptance through DVT. Ideally, it is hoped that the product will function beyond the Upper and Lower Spec limits, this is commonly

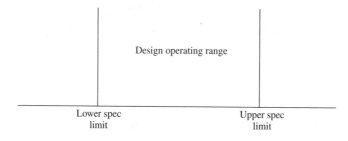

Design operating range

Lower spec
limit

Upper spec
limit

Figure 7.14   Product design specification limits

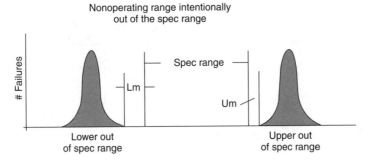

Figure 7.15 Design margin

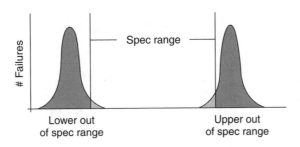

Figure 7.16 Some products fail product spec

referred to as *design margin*. This "design margin" provides a safety range that allows for component, design and process drift that would otherwise reduce the yield of the product in production. Over time, in the field, product performance begins to drift, leading to system failure. Design margin helps maintain the product operation over time. This point is illustrated in Figure 7.15. The shaded curves show the distribution of where a sample of products fails. The graph shows that when the sample is stressed to the limits of the specification there are no failures. In fact, the first product failures begin to occur at a point beyond the upper and lower design margins. You may be wondering why there is no shaded coned distribution beyond the upper and lower design limits. To explain this, consider that we are going to stress test 1,000 production to find the upper and lower limits where the product fails. What we will find is that they will not all fail at the very same point. There will be one point where most of the units fail. Then, as we go above and below that center point, we will find the number of units that fail decrease and eventually go to zero.

Upon testing the first prototypes, often, not all the design specifications are met. Sometimes, there is little or no margin for safety. Figure 7.16 illustrates a specification where the upper and lower "out of spec ranges" have fallen inside the design spec range. In this situation, not all the products are able to meet the design specification. This problem will manifest itself in manufacturing as a low first pass yield and early field failures.

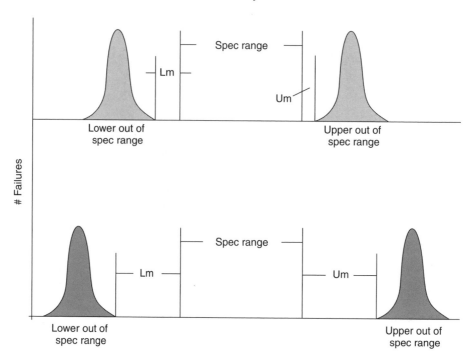

Figure 7.17 HALT increases design margin

HALT testing will improve product reliability, DVT acceptance and first pass production yields by increasing the design safety margins. To illustrate this point, refer to Figure 7.17.

In HALT testing, we stress the product beyond its design specifications. At some point, the stress becomes so great that the product no longer operates. This is referred to as *a failure*. However, there are two types of failures that are possible, these are called *soft* and *hard failures* as shown in Figure 7.18. They are sometimes referred to as *recoverable* and *nonrecoverable failures*. To determine which type of failure you have, reduce the stress level to its normal specification range. If the product returns to normal operation, then it is a soft failure. If the product still does not operate, it is a hard failure. The product may need to be reset before it can return to normal operation. If so, this is still considered a soft failure because no rework was required to fix the product. Hard failures require troubleshooting to determine what failed. All hard failures are later investigated to determine the root cause of the failure.

In HALT testing, the product is stressed to hard failure, the root cause of the failure is determined and appropriate design changes are implemented. The soft failures are also designed out. To verify that the design change has been successful, we perform a second HALT test. If the failure modes are removed

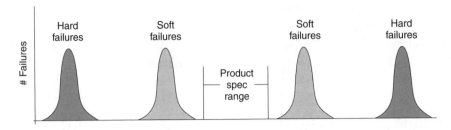

Figure 7.18 Soft and hard failures

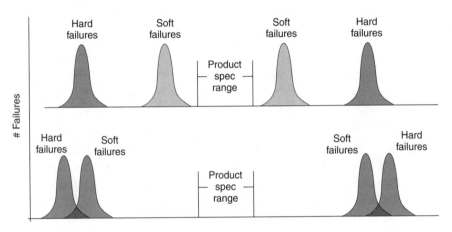

Figure 7.19 Impact of HALT on design margins

and no new failure modes surface, then design fix is considered good. So how does HALT improve product reliability, DVT acceptance and first pass production yields?

The hard and soft failures that were precipitated in HALT are designed out of the product. In doing so, it increases the "design margin" where the stress causes product failure. Fixing the hard and soft failures causes the design margins to widen. By correcting the failures found in HALT, the gray shaded areas in Figure 7.17 are pushed out, leaving greater design margin between the design specifications and the stress points where failures occur. The end result is an improved product reliability and improved production yield due to improved sensitivity to process variation.

HALT intentionally stressed the product beyond the spec limits in order to cause failures. Then, the failures are corrected. The resultant product is more reliable, because the field failures surfaced in design and were fixed. Also, the product can now operate to specification limits beyond the design specifications. This point is further illustrated in Figure 7.19.

The Product Spec Range lies between the upper and lower operating points where the product will remain in specification. Ideally, the product will have some margin where it will still function without failure. The measure of this is the Upper and Lower Operating Margin. When stresses are applied to the product beyond these margins, then soft failures will start to occur. Continuing to increase the stresses will drive the product to hard failures.

7.2.3 HALT Testing

Before the product can be HALT tested, some planning is in order. There can be a significant amount of preparation work required before HALT testing. The reliability engineer and the lead engineer should work together to have everything in place for the day the HALT process is to begin. The following is a list of items that should be ready for the test:

1. The product, hopefully, five working units, and a spare. The spare is often called a *gold* unit because it is not intended for stress testing; it is used when there are subtle testing issues and it is difficult to tell if the DUT or the test instrumentation is at fault. Inserting the gold unit will verify if the problem exists with the DUT. This can speed the troubleshooting process greatly. Thus, the information learned by using this unit is "golden."

2. Test instrumentation. This is probably the most important item on the list after the product itself because the failures have to be discovered and corrected. Poor monitoring will miss some failures and render the HALT process less effective than it might have been.

3. The output specifications that will be monitored and the monitoring instrumentation.

4. Documentation, that is, schematics, assembly drawings, flowcharts, and so on.

5. A mechanical fixture to affix the DUT to the HALT table.

6. Input and output cabling.

7. Special devices, that is, liquid cooling apparatus, air ducts, power sources, other support devices, and so on.

8. Software, where required.

9. The stress levels intended to be applied to the DUT (established and agreed to by the HALT team).

10. The time required for the testing.

11. The lead engineer needs to be scheduled for the entire HALT process.

12. A test engineer to assist in failure analysis.

13. The reliability engineer and a HALT chamber operator. (The reliability engineer usually writes the final HALT test report.)

The HALT team starts by placing the DUT in the HALT chamber and interconnecting it to the power sources, loads and instrumentation. Then, the

DUT is turned on and monitored to verify if it is operating properly. This step is just to make sure the new set up is functioning properly. The runtime here is determined by how long it will take to verify that the unit is functioning properly. It is often a function of the time it takes the test software to run one or two complete test cycles to completion.

Then the HALT chamber doors are closed and a low level of vibration is started. This should be in the range from 2 to 4 Grms, random vibration over 6 degrees of freedom. The purpose is to verify that the interconnections and monitoring are hooked up properly and there are no loose connections. Again, one to two test cycles are run to verify that the hardware is ready for the HALT.

Next, the chamber should be driven to low temperatures in 5 to 106 °C steps with dwell times of typically 10 min. (Here the idea is to use the weakest stress and move to stronger stresses as the testing continues. This way, the subtle failures will not be lost with excessive stress testing.) The dwell times are sometimes driven by the test instrumentation time required to complete a full test. Continue in steps to the cold limit and complete one more dwell period. Then, return to room temperature. Begin to increase the temperature in a similar manner to the high temperature limit. Record all failures. Stop in mid test to analyze failures and see if that can be driven to root cause. If, and when failures occur, see if the failure can be found while still in the fixturing. There may be the possibility to "band-aid" the failing element in order to continue increasing the stress. After the high temperature has been reached, complete one more dwell period and return to room temperature. The first stress element, temperature, is complete.

Many practitioners then move to rapid thermal stress testing. This is where the chamber temperature is made to change as rapidly as possible. The temperature levels should be 5% below the high and low temperature extremes used in the step testing. This test method uncovers the extreme thermal rate of change weaknesses. Run several rapid temperature excursions – three to five cycles will suffice.

Then, vibration is applied to the DUT. Increase vibration by 5 to 10 Grms levels until you reach the limit of the chamber's capability or the DUT is nearing destruction. However, when 20 Grms is reached, lower the vibration to 1 to 2 Grms (tickle vibration) for one test cycle. Many times, vibration-caused failures do not reveal themselves to the test instrumentation at the higher vibration levels, but the failure becomes apparent at the lower levels. Then, for every step, increase in vibration, dwell and test, and return to the low vibration level again. Continue until the highest vibrations that were established for the test are reached. Record the failures and of course, troubleshoot to the root cause.

Now combine the temperature and vibration stresses. Work the chamber stresses simultaneously in steps, as before. Then, perhaps power supply voltage margining can be added, first alone, and then added to the first two stresses.

Other stresses can be combined as well (AC line input voltage and frequency margining, timing margining, etc.).

Repeat these stresses in the same sequence on the next DUT and carefully note where there are similarities and differences. Continue until all the DUTs have received HALT and the data is recorded. It might be best, in some cases, to save some of the DUTs and stop testing early. This will allow you to work on the failures and the root causes, and to apply the fixes to new systems.

It is very important to record the stress levels where the soft and hard failures occurred. Later, when you have made design corrections these stress levels should have increased, thus increasing your Operating Margins.

A relatively new HALT technique created by Dr. Greg Hobbs is the "search pattern technique." The idea is to slowly sweep temperature and rapidly sweep vibration simultaneously. Starting with the product at room temperature (or about 25 °C), the temperature is lowered to the lower stress limit, say, −40 °C. At the same time, vibration is sweeping as fast as it can between 0 to 20 Grms. Typically, the vibration will go from the low-level to the high-level and back down again in less than 30 s (this is adjustable on some HALT chambers). Once the vibration stresses are started, the temperature is slowly swept from −40 °C to +140 °C (hypothetical values) and then back to room temperature. If the temperature rate of change is set to 2 °C/min, the entire test will take 4.4 h (refer to Figure 7.20).

The search pattern technique is valuable where the soft failure is very close to the hard failure. The temperature changes slowly while the product is being

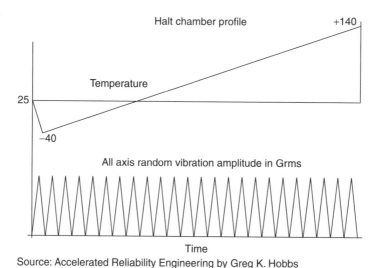

Figure 7.20   HALT search pattern

continuously monitored. This allows the test to be stopped before a hard failure is encountered. This opens opportunities for some failure investigation before they occur, only at a specific temperature. If you use step stresses, there is a possibility that you will pass over the point of instability.

7.3 HIGHLY ACCELERATED STRESS SCREENING (HASS)

Once the reliability design issues identified from HALT testing have been designed out, the product is ready for manufacturing. However, a good design is only half the battle. Product reliability is achieved through a good design and manufacturing process. Contrarily, design flaws and poor manufacturing processes result in field failures. The HALT process focused on the design issues that result in field reliability issues. There is also manufacturing process variation, which can produce product weaknesses that eventually lead to field failures. Can the HALT process be applied to manufacturing to prevent products from shipping which have an unacceptable process variation?

The HALT process can be applied indirectly to the manufacturing process. The process that is used in manufacturing is called *HASS* (*Highly Accelerated Stress Screening*). HASS can prevent marginal and defective units from being sold. The purpose of HASS is to identify products that have process-related defects and manufacturing weaknesses before shipment. HASS is also helpful in identifying when suppliers provide potentially defective parts.

HALT and HASS use similar types of accelerated stresses to identify failures. Both processes also stress the product when it is powered up and operational. But the similarities stop there. HASS stresses the product in a similar way to HALT but at reduced stress levels. HASS is a gentler form of HALT. HALT is a means to stress test the design and we emphasize the "T" for test. HALT is intended to reveal design-related failures. It is a proactive tool to improve product design and is performed by design engineering. HALT testing reveals failures that customers will experience in the field if the design is not fixed. HASS is a stress screen with the last "S" used to emphasize screen. After products are manufactured, they pass through the HASS screen to verify that the manufacturing process is in control. The HASS test is performed in manufacturing as a means of product acceptance. Process variations are flagged by HASS and can immediately be corrected to prevent an unsatisfactory product from being shipped. HASS is a reactive tool to assure that the manufacturing process stays in control.

The more complex the manufacturing process, the greater the opportunity for manufacturing defects to enter the product, rendering it less reliable. Control of the manufacturing process is critical. Even with the best manufacturing practices in place, that is, Statistical Process Control (SPC), continuous improvement, Electrostatic Discharge (ESD) protection and training, manufacturing defects due to process drift and supplier issues can surface. The HASS screen is intended to either pass or "screen out" nonconforming product. A HALT chamber, added to the end of the production line, which applies reduced stresses to the product can perform the vital HASS screen. The HALT chamber and the HASS chamber can be the same. This is an option for small companies who outsource manufacturing or produce products in low volumes. Many companies prefer to avoid the scheduling conflicts between design and manufacturing by having separate chambers for HALT and HASS. HASS chambers also tend to be bigger for batch processing.

From the HALT test, the product operating and destruct limits are learned. In the HASS screen, the product is stressed at levels beyond its operating limits but below the destruct limits. The HASS profile consists of two parts, the precipitation screen and the detection screen. The test begins with the precipitation screen. The precipitation screen is a stress level that is below the destruct limit and above the operating limit. Refer to Figure 7.21. The HASS screening level applied to the product needs to be determined. A good stress level for temperature is between 80% and 50% of the destruct limits. The initial vibration stress level is set at 50% of destruct limits. It is important to stay below the destruct limits; otherwise damage to good product is likely. The purpose of the precipitation screen is to sufficiently damage defective products so they can be detected later in test. However, the stress must not damage or severely degrade good product. Generally, if the right stress levels are applied, the defective assemblies will degrade at a significantly greater rate than good product. A Proof Of Screen (POS) (discussed in Section 7.3.1), will identify if the precipitation stress is too severe or ineffective.

The precipitation stress was designed to sufficiently damage defective product so it can be differentiated from good product. The way we identify bad products is through a detection screen. The detection screen applies temperature stress

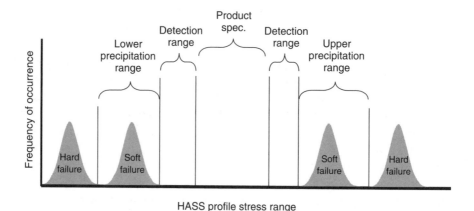

Figure 7.21   HASS stress levels

at levels that are between the soft failure limit and the product spec limit. Set the temperature stress midway between the spec limit and the soft failure limit. The vibration level is set between 3 and 5 Grms (often referred to as *a tickle vibration*). The HASS profile is usually short, typically 3 to 5 cycles of precipitation and detection is adequate.

7.3.1   Proof Of Screen (POS)

The environmental stresses induced on the product by the HASS screen will remove some of the life expectancy of the product. This is unavoidable. The goal of the HASS screen is to provide a stress level high enough to precipitate identification of manufacturing defects without removing an excessive amount of product life. How much product life is removed in HASS can be estimated through a process called *Proof Of Screen* (POS).

The POS process is simple; just repeat the HASS screen until the product fails. Applying the HASS stress repeatedly causes the product to degrade at an accelerated rate. Eventually, the product will fail because of the accumulated effect of the stress. If it takes 20 times to render the product nonoperational, then it is reasonable to estimate that 5% off the product life is removed with each HASS screen. If the DUT failed after only four HASS screens, it can be assumed that 25% of the life of the product was removed each time. There is no minimum number of stress cycles desired before a product fails. Some companies want at least 20 cycles without a failure. The test should be run on a large enough sample to assure that normal manufacturing process variation is accounted for.

If, on the other hand, you run the HASS test for 100 cycles without a failure, the HASS stress levels may be set too low. Some practitioners recommend seeding product to determine if the HASS screen is effective at detecting defective product. Seeding a board requires intentionally inserted manufacturing defects into the product. The product is then tested to determine if the defects are found during HASS screen. The problem with seeded defects is that it is difficult to insert seeded defects that are real representations of product defects (i.e., manufacturing process drift or supplier changes).

Are there alternatives to HASS? Is HASS the only way to screen manufactured products for defects? No, there are other techniques but HASS is the most effective. The alternatives are Burn-In, Environmental Stress Screening (ESS), and of course electrical test with no environmental stresses at all.

7.3.2   Burn-in

The burn-in process can be applied to components and final products as a final acceptance test. Recall the popular bathtub curve from Chapter 6; Figure 7.22. The early failure rate of a product is often higher than the failure rate during its

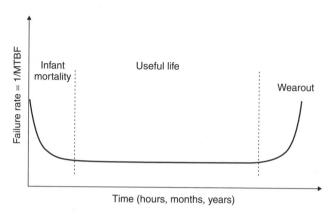

Figure 7.22 The bathtub curve

useful life. It is the reason some people prefer to buy a year-old car because "the bugs have been worked out." Burn-in is designed to accelerate infant mortality so failures occur before the product is sold. Typically, infant mortality failures occur in the first year of product use. Product failures are typically higher in the first year, when reliability and quality problems often surface. Theoretically, you could avoid the high infant mortality failure rates by operating the product in-house for a year before it is sold. This obviously is impractical for many reasons. However, if you can accelerate the products first year use, that is, "burn-in the product," then the infant mortality failure rate will occur during the burn-in test. The burn-in must be long enough to remove most early life failures.

The burn-in process is performed on 100% of production and usually consists of powering on and off the product while running a test diagnostics. It is common for a burn-in test to run 24 to 48 h. The product is also kept at an elevated temperature typically at the upper limit of its product specification range, through a temperature chamber.

The burn-in process typically takes, one- to two-days test time, but it can be longer. To get around this long test time, burn-in test chambers tend to be large so that many production units can be tested concurrently. This lowers test costs and increases throughput. Large burn-in chambers can cost several hundred thousand dollars.

The burn-in test, sometimes referred to as a *biased bake test*, is designed to accelerate the aging process. In the 1970s and 1980s, it was not uncommon for products to have a high infant mortality failure rate because the components they used had considerable variation in them. The advent of concurrent engineering, design guidelines, and quality programs such as Total Quality Manufacturing (TQM), continuous improvement and SPC component quality has changed all that. Components, today, have increased in reliability by orders of magnitude. Today, almost all component manufacturers deliver reliable parts. The reliability of the product is no longer driven by the quality of the components, but by the quality of the design and manufacturing process. This does not mean that part selection is no longer an issue. If you select the wrong part for the job, then expect to have a reliability problem. But the problem is no longer "bad parts."

In addition, studies have shown that burn-in is not an effective technique to remove early failures. Research indicated that burn-in at the component level tended to damage more good parts (via ESD, handling and electrical overstress) than identify bad parts.

7.3.3   Environmental Stress Screening (ESS)

Another common form of a "burn-in" test is called *Environmental Stress Screening (ESS)*. An ESS test is an environmental stress test designed to accelerate the failure of faulty product. The test is performed while the product is operational and being monitored. The main difference between ESS and conventional burn-in is that ESS induces multiple stresses on the product. These stresses might include

- temperature cycling,
- temperature soak,
- vibration.

However, unlike HASS, the stresses applied to the product are generally below the product spec limits.

Conventional ESS testing can be as short as 2 to 4 h or as long as 24 h depending on the test. Completed products are placed into an ESS chamber (temperature/vibration chamber) and the power is turned on. A typical ESS starts with a temperature cycling profile where the temperature is increased to just below the design specification of the product. When the upper temperature is reached, the product is temperature soaked from 1 to 2 h. Monitoring equipment will detect if the product goes out of specification. After a high temperature soak, the product is transitioned to low temperature while still biased and operating. When the lower operating temperature is reached, the product is again cold soaked 1 to 2 h. After the temperature soak portion of ESS, a rapid temperature cycling test is performed. The temperature is raised and lowered in repetitive cycles while monitoring continues.

If through this sequential process the ESS total time is reduced to 1 day, then manufacturing process-related failures occurring beyond 1 day would probably not be detected if the process goes out of control again. Another shortcoming of the ESS burn-in process is that it may be a week before a manufacturing process error is discovered. This means that the manufacturer could produce many products that are not conforming and still need to be corrected.

This is a sort of a Catch-22. Short ESS cycles are desired to maintain low work in process costs and to reduce the size and cost of the ESS resources needed. Long ESS cycles are desired to catch manufacturing process defects that are undetectable with short ESS cycles. Reducing the ESS window is a large risk. It is easy to see that the ESS concept can provide a poor compromise.

Solder failures and connecting lead technology are high on the list of failures discovered in the field. The majority of these failures that can be detected by temperature cycling cannot be detected with these few cycles, even in a 5-day ESS. This is why the customer discovers these failures after several months or years in the field. Often, several thousand cycles are required to precipitate this type of failure. You cannot afford to use the ESS process to discover these failures.

7.3.4   Economic Impact of HASS

HASS accommodates few units, simply because the chambers tend to be relatively small. (Some HASS chamber manufacturers will provide customized chambers for the specific needs of the manufacturer.) If the HASS process uses rapid temperature cycling, then the number of units in the chamber will necessarily have to be limited so that the internal temperatures of all the DUTs can be achieved rapidly. Because HASS can often be accomplished in a short time, such as 30 min to 60 min, a high number of units can pass through the HASS process daily. Considering that the HASS process can also look at a smaller sample, it can be much closer in real time to when a defect might have been inserted into the process. This translates into a quicker recovery process when using HASS. Because most defects found in manufacturing are process-related defects and not design-related ones, process delay due to detection and improvement is considerably shorter. Once the HASS process has the manufacturing process in control, the cost of ensuring top-quality is significantly reduced.

The manufacturing ramp means that few units are produced at first, more are produced later and then full production volumes are reached. HASS needs to be applied to every unit until it is clear that there are no new process-related failures. It may be unnecessary to continue HASS because the process is in control, largely due to HASS discoveries. But it would be unwise to completely curtail HASS because process drift and change creep into the system with negative reliability impact. An audit process should be incorporated.

7.3.5   The HASA Process

Highly Accelerated Stress Audit™ (HASA) is a HASS audit process. It periodically examines the process and adds no significant cost when it is not auditing.

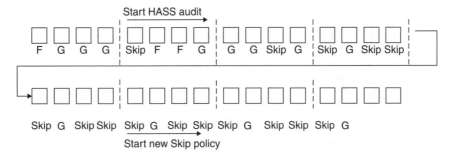

Figure 7.23 HASA plan. Courtesy of James McLinn

It is reasonable to assume that all the manufacturing processes will not stay in control forever, but only for a while. Once HASS has verified the manufacturing process is under process control, the screening can be moved to a skip-lot audit process. The number of units that can be manufactured without HASS has to be determined by the nature of the manufacturing process. If 100 units can be produced everyday and placed in shipping hold, the HASS process can be an audit. Divide the 100 units into 4 groups of 25, sample the groups with the following results shown in Figure 7.23.

The first lot has a failure and the next 3 lots are all good. Skip the next lot and sample the one after. This has a failure, go back and check the skipped lot. Next, sample the next two. The next lot has a failure and the one after is good. Sample the next two and both are good. Skip the next lot and sample the one after. This is good so skip the next lot and sample the one after. It is also found to be good. So skip the next three lots. Sample the one after; it is good so continue to skip the next three lots. Continue in this fashion until there is a failure and go back to every other lot until you find three good samples in a row. This is a standard skip-lot process.

As you can see, the number of units to be produced between HASS audits (HASA) is a function of many variables. Ideally, the number of units produced between HASA screens should be as high as possible and yet small enough to accommodate cost-effective correction action.

7.4 SUMMARY OF HALT, HASS, HASA AND POS BENEFITS

1. Electronic design/margin improvement
2. Packaging design improvement
3. Parts selection improvement
4. Production process improvement
5. Software implementation improvement

6. Rapid design and process maturation

7. Reduced total engineering time and cost

8. Lowered warranty costs

9. Higher mean time between failures

10. Rapid process corrective action.

7.5   HALT AND HASS TEST CHAMBERS

A brief description of the HALT chamber is in order. First, let's describe what it is not with a description of the typical burn-in type process performed by many manufacturers. Typical environmental chambers have the ability to raise and lower the temperature of the chamber interior. The chamber is heated by applying current through resistance wires. Cooling is usually accomplished using a form of air conditioner. (Some of these temperature chambers use liquid nitrogen.) The temperature of the oven can be increased rapidly using resistance wires but it cannot be cooled rapidly. The air conditioner cooler system does not have the thermal capability of lowering the temperature in the chamber rapidly. This shortcoming of standard temperature control for environmental chambers leads to long test cycle times. A single temperature cycle from ambient to 140 °C and back down to −40 °C with dwells of 1 h at the high and low temperatures may take 5 or 6 h to complete owing to the slowness of the air conditioner and the thermal mass of the devices in the chamber. This inability of standard burn-in ovens makes temperature cycling of production units undesirable to the manufacturer.

HALT chambers are environmental chambers designed to quickly provide two environmental stresses. Typical HALT chambers can control temperatures from −100 °C to +200 °C. By using huge resistor banks and a specially tuned liquid nitrogen cryogenic system, these chambers can produce temperature rates of change of 60 °C/min. Some chambers with advanced cryogenic management can achieve rates of change in the 80 °C/min range. Through a special design, liquid nitrogen can be aspirated into the chamber very rapidly and the device in the chamber will respond much faster than it can in the burn-in type chamber. In the HALT chamber, a multitude of vents and hoses direct high-flow cooling or heating air toward the device in test to achieve very rapid temperature rates of change of the device in test. Thermocouples located inside the DUTs will ensure that the internal temperatures are achieved quickly to minimize overall cycle time.

The vibration table in the HALT chamber is also unique. It has the ability to move in three directions linearly and rotationally; hence the term *six degrees of freedom* (refer to Figure 7.24).

There are compressed air-driven piston actuators mounted under the table in many angles and directions. These actuators operate in a random sequence.

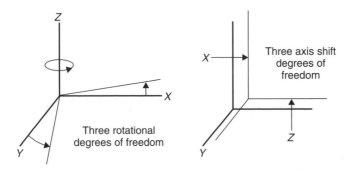

Figure 7.24 A HALT chamber has six simultaneous degrees of freedom (movement)

They impart their energy to the table; thus, the table moves in harmony with the actuators. Control electronics randomly selects the actuators and controls when they operate and to what magnitude. It is easy to imagine that the underside of the table has 8 to 16 miniature air hammers mounted in all directions that are operated by miniature operators running randomly all at the same time. The table is mounted on a cushion and springs so that it can move in six degrees of freedom. Test devices mounted to the table move similarly. The frequency response of the actuators imparts a broad frequency range to the device under test. Typical table vibration frequencies range from 2 Hz to 10,000 Hz. Tables typically can produce vibration levels to upwards of 60 Grms to devices in test that reach 1,000 pounds or more.

7.6 SPC TOOL

Reliability is "Quality over Time." Early in this book, we discussed the difference between reliability and quality; here, we will point out how to use a well-understood quality tool to improve reliability. Statistical Process Control (SPC) is a tool used in manufacturing to minimize control process drift. The way the tool is used is to periodically monitor a given process to ensure that a product parameter has not drifted out of specification. The SPC process establishes high and low levels that are not to be exceeded. When the manufacturing process drifts near these limits, the operator is instructed to make an adjustment to return the manufacturing process to the center of these limits or to an ideal setting so that quality and consistency can be maintained as shown in Figure 7.25.

Consider that the entire manufacturing processes may be under SPC and at any given time many of the monitored processes have drifted too near the high level. If the finished product is to be assembled using all the processes at this time, this particular unit may still be within the quality standard, still conforming to all specifications, but is on the verge of falling out of tolerance in all the areas being monitored by the SPC process. Compare this unit to one that

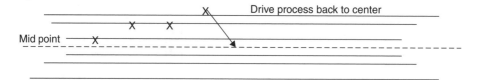

Figure 7.25 Selective process control. Courtesy of James McLinn

was manufactured when all the SPC controls were near their midpoints. This latter unit will obviously perform within the design specifications for a long time before any of these parameters go out of tolerance from normal process drift or accumulated wear. The former unit is on the verge of being out of quality specification initially, and very little stress and use may drive it outside its design specifications. This unit has a low reliability and may be caught by the HASA test. Wider design margins translate into improved reliability in the field.

Consider two units that meet the initial quality specification, but only one unit has high reliability because it was manufactured close to nominal tolerances and the other was not. The unit farther from nominal will cost more warranty dollars. This is the basis of the Taguchi loss model. The challenge is to identify the important few limits that can economically be controlled in the manufacturing process so that units stay well within acceptable quality and reliability standards. The SPC tool has been used to produce products to specification, and by tightening on the upper and lower tolerances for the critical few processes, you can produce products that also have greater reliability. When used as described above, SPC can be a reliability tool as well as a quality tool.

7.7 FIFO TOOL

Rotating inventory so that units on the shelf the longest get used first is a well-established business tool. The accounting term, FIFO, "First In, First Out," can apply to material handling to improve reliability as well. Electrical components have leads that will eventually be connected together to make a larger assembly. Usually, this means they will be attached to circuit boards, either by surface mount, plated-through-hole technology or by the surface-to-surface (connectors) mating. The longer these components remain in inventory, on the shelf, the likelier it is that the electrical leads of these components will begin to oxidize and corrode. The solderability of these leads is greatly reduced by this oxidization process. The typical electronic assembly has hundreds and often thousands of electrical connections. If a small number of these connections are made unreliable by using components with questionable leads the overall reliability of the assembly may be reduced.

One way to reduce lead oxidation is by placing all the components in bags or containers that are filled with nitrogen gas. Component lead degradation will be greatly reduced by reducing the oxygen environment around components. As part of the purchasing process, materials that are susceptible to oxidization can be purchased in nitrogen filled containers. This is more expensive than using FIFO to control inventory.

By not using FIFO, quantities of older components and material collect in the back of stockroom shelves. Recently purchased material is placed on the front of the shelf and is often the first material used in manufacturing. There will be times when, to fill orders, all the material on the stock shelf will be required. In this case, there is a high likelihood that there will be solder reliability problems in manufacturing and later in the field. Solder and connection reliability can be maintained at high levels by controlling order quantities and using the oldest material in the stockroom. Remember also that you don't know how long those components may have sat on your distributor's shelves before they were shipped to your company.

Integrated circuits are also a critical component. They have leads that are susceptible to oxidation like most other components, but they have another problem. Plastic encapsulated Integrated Circuits absorb water from the atmosphere while they are waiting to be soldered into a final assembly. This can be a real problem in manufacturing. When a component goes through the soldering process, it is often heated to temperatures well above the boiling point of water. The moisture in the component changes to steam. The pressure inside the component from the steam is sometimes great enough to cause microcracks and even cause delamination. Many times, these potential failures will not be discovered during test. They will often manifest themselves as early field failures and high warranty costs. This failure mechanism is called *the popcorn effect* because the failure is caused just like popcorn.

It is recommended that integrated circuits and other plastic devices that are susceptible to the popcorn effect be purchased and stored in nitrogen containers. Additionally, these components should be preconditioned in a warm baking oven for 24 to 48 h prior to passing through the soldering process. This preconditioning causes the moisture that has collected inside the integrated circuit to slowly migrate out. Thus, the popcorn effect is eliminated by process changes.

Sometimes, some or all the material selected for manufacturing may not be needed and may be stored to be used another day. It should be returned to nitrogen containers or placed in process so it can be preconditioned again before being placed through the solder process. Unused material often gets set outside the standard manufacturing loop so attention to this detail is important. The amount of time between completion of preconditioning and placement in the solder process varies depending on the components and the local environmental conditions. The desired temperature for the preconditioning oven varies but is obtainable from the component manufacturer.

7.8   COMPONENT DERATING – A FIRST LINE OF DEFENSE IN PRODUCT RELIABILITY

Earlier, we stated that the reliability of most components has improved 10 to 100 times in the last two decades. Some claim examples, showing as much as a ten thousand-fold improvement. Improvements of four orders of magnitude may be true in some cases, but using components properly and not overstressing them will pay large dividends in terms of system reliability. This will make components a very small part of the unreliability picture.

A design practice that selects parts to be stressed (in circuit) to a value well under their individual rated limits is called *derating*. An important specification for capacitors is breakdown voltage. It is well advised to select capacitors that have voltage and temperature ratings well above the specific needs of the circuit design. Diodes may have a peak reverse voltage rating. Selecting diodes that have a rating well above that of the circuit application will lower the stress on the diode and it will perform longer in the application. There are so many different electronic components that they cannot be enumerated here. The three components noted here often have several parameters that should be derated, not just those mentioned. Examples of such derating systems include Mil-Std 975G.

Derating can be seen everywhere. Sometimes it is for public safety. The maximum weight requirements for an elevator are one form of derating.

In the simplest terms, it means that if a component part, such as a resistor, can dissipate 1 W, use it in a circuit that never requires more then half of that, it will withstand the stresses of the circuit longer. Many components are subject to accumulative fatigue due to applied stresses. Derating reduces the impact of these stresses and greatly extends the life of the component when derating is also applied to the load. The performance may improve as well.

In all the engineering specialties, there are specifications for parts of every sort. Valves have pressure limits, cables have load limits, and materials have temperature limits. Design these and all components such that, by design, they are not the critical part of the reliability picture.

References

1. R. J. Geckle, R. S. Mroczkowski, *Corrosion of Precious Metal Plated Copper Alloys Due to Mixed Flowing Gas Exposure*, Proc. ICEC and 36th IEEE Holm Conference on Electrical Contacts, Quebec, Canada, 1990, and IEEE Trans. CHMT (1992).
2. R. Gore, R. Witska, J. Ray Kirby, J. Chao, *Corrosive Gas Environmental Testing for Electrical Contacts*, IBM Corporation, Research Triangle Park, NC.

FMEA

1. *Recommended Failure Modes and Affects Analysis (FMEA) Practices for Non-Automobile Applications*, SAE (2001).

2. M. Krasich, *Use of Fault Tree Analysis for Evaluation of System Reliability Improvements in Design Phase*, 2000 Proceedings Annual Reliability and Maintainability Symposium (2000).
3. K. Onodera, *Effective Techniques of FMEA at Each Life-Cycle Stage*, 1997 Proceedings Annual Reliability and Maintainability Symposium, IEEE (2000).
4. S. Bednarz, D. Marriot, *Efficient Analysis for FMEA*, 1998 Proceedings Annual Reliability and Maintainability Symposium (1998).
5. M. Kennedy, *Failure Modes and Effects Analysis (FMEA) of Flip-Chip Devices Attached to Printed Wiring Boards (PWB)*, IEEE/CPMT International Manufacturing Technology Symposium, IEEE (1998).
6. R. Whitcomb, M. Riox, *Failure Modes and Effects Analysis (FMEA) System Development in a Semiconductor Manufacturing Environment*, IEEE/SEMI Advanced Semiconductor Manufacturing Conference, IEEE (1994).
7. D. J. Russomanno, R. D. Bonnell, J. B. Bowles, *Functional Reasoning in a Failure Modes and Effects Analysis (FMEA) Expert System*, 1993 Proceedings Annual Reliability and Maintainability Symposium, IEEE (1993).
8. S. Prasad, Improving Manufacturing Reliability in IC Package Assembly Using the FMEA Technique, *IEEE Transactions of Components, Hybrids and Manufacturing Technology*, **14**(3), 452–456 (1991).

HALT

1. General Motors Worldwide Engineering Standards, *Highly Accelerated Life Testing*, GM (2002).
2. G. K. Hobbs, *Accelerated Reliability Engineering*, John Wiley & Sons (2000).
3. H. W. McLean, *HALT, HASS & HASA Explained: Accelerated Reliability Techniques*, American Society for Quality (May, 2000).
4. J. Strock, Product Testing in the Fast Lane, *Evaluation Engineering*, (March, 2000).
5. N. Doertenbach, *High Accelerated Life Testing – Testing With a Different Purpose*, IEST, 2000 proceedings (February, 2000).
6. D. Rahe, *The HASS Development Process*, 2000 Proceedings Annual Reliability and Maintainability Symposium, IEEE (2000).
7. D. Rahe, *The HASS Development Process*, ITC International Test Conference, IEEE (1999).
8. M. Silverman, *Summary of HALT and HASS Results at an Accelerated Reliability Test Center*, Qualmark Corporation, Santa Clara, CA, 1998 Proceedings Annual Reliability and Maintainability Symposium, IEEE (1998).
9. M. Silverman, *HASS Development Method: Screen Development, Change Schedule, and Re-Prove Schedule*, 1998 Proceedings Annual Reliability and Maintainability Symposium, IEEE (1998).
10. R. H. Gusciaoa, *The Use of Halt to Improve Computer Reliability for Point of Sale Equipment*, 1998 Proceedings Annual Reliability and Maintainability Symposium, IEEE (1998).
11. J. A Anderson, M. N. Polkinghome, *Application of HALT and HASS Techniques in an Advanced Factory Environment*, 5th International Conference on Factory 2000, (April, 1997).

12. M. L. Morelli, *Effectiveness of HALT and HASS*, Hobbs Engineering Symposium, Otis Elevator Company (1996).
13. C. Ascarrunz, *HALT: Bridging the Gap Between Theory and Practice*, International test Conference 1994, IEEE (1994).
14. R. Confer, J. Canner, T. Trostle, S. Kurz, *Use of Highly Accelerated Life Test Halt to Determine Reliability of Multilayer Ceramic Capacitors*, IEEE (1991).
15. H. McLean, *Highly Accelerated Stressing of Products with Very Low Failure Rates*, Hewlett Packard Co., (1991).
16. P. E. Joseph Capitano, Explaining Accelerated Aging, *Evaluation Engineering*, p. 46 (May, 1998).
17. E. R. Hnatek, Let HALT Improve Your Product, *Evaluation Engineering*.
18. G. K. Hobbs, What HALT and HASS Can Do for Your Products, *Evaluation Engineering*.
19. G. K. Hobbs, What HALT and HASS Can Do for Your Products, *Hobbs Engineering, Evaluation Engineering*, Qualmark Corporation, p. 138 (November, 1997).
20. E. O. Minor, *Quality Maturity Earlier for the Boeing 777 Avionics*, The Boeing Company.
21. M. A. Silverman, *HALT and HASS on the Voicememo II™*, Qualmark Corporation.
22. M. Silverman, *Summary of HALT and HASS Results at an Accelerated Reliability Test Center*, Qualmark Corporation, Santa Clara, CA.
23. M. Silverman, *Why HALT Cannot Produce a Meaningful MTBF Number and Why this should not be a Concern*, Qualmark Corporation, ARTC Diivision, Santa Clara, CA.
24. W. Tustin, K. Gray, Don't Let the Cost of HALT Stop You, *Evaluation Engineering*.

HASS

1. T. Lecklider, How to Avoid Stress Screening, *Evaluation Engineering*, pp. 36–44 (2001).
2. D. Rahe, *The HASS Development Process*, 2000 Proceedings Annual Reliability and Maintainability Symposium, IEEE (2000).
3. M. Silverman, *HASS Development Method: Screen Development, Change Schedule, and Re-Prove Schedule*, 2000 Proceedings Annual Reliability and Maintainability Symposium, IEEE (2000).
4. D. Rahe, *HASS from Concept to Completion*, Qualmark Corporation.

Quality

1. S. M. Nassar, R. Barnett, *IBM Personal Systems Group Applications and Results of Reliability and Quality Programs*, 2000 Proceedings Annual Reliability and Maintainability Symposium (2000).
2. D. K. Ward, A Formula for Quality: DFM + PQM = Single Digit PPM, *Advanced Packaging* (June/July, 1999).

3. S.-B. Lee, A. Katz, C. Hillman, Getting the Quality and Reliability Terminology Straight, *IEEE Transactions on Components, Packaging, and Manufacturing*, **21**(3), 521–523 (1998).
4. Carolyn Johnson, Before You Apply SPC, Identify Your Problems, *Contract Manufacturing* (May, 1997).
5. C.-H. Mangin, *The DPMO: Measuring Process Performance for World-Class Quality*, SMT (February, 1996).
6. H. L. Oh, A Changing Paradigm in Quality, *IEEE Transactions On Reliability*, **44**(2), 265–270 (1995).
7. T. A. Pearson, P. G. Stein, On-Line SPC for Assembly, *Circuits Assembly*, (October, 1992).
8. G. Kelly, SPC: Another View, *Surface Mount Technology* (October 1992).
9. P. Gupta, Process Quality Improvement – A Systematic Approach, *Surface Mount Technology* (August, 1992).
10. E. O. Minor, *Quality Maturity Earlier for the Boeing 777 Avionics*, The Boeing Company.

Burn-in

1. D. R. Conti, J. Van Horn, *Wafer Level Burn-In*, Electronic Components and Technology Conference, IEEE (2000).
2. J. Forster, *Single Chip Test and Burn-In*, Electronic Components and Technology Conference, IEEE (2000).
3. T. Sdudo, *An Overview of MCM/KGD Development Activities in Japan*, Electronic Components and Technology Conference, IEEE (2000).
4. C. F. Hawkins, J. Segura, J. Soden, T. Dellin, Test and Reliability: Partners in IC Manufacturing, Part 2, *IEEE Design & Test of Computers*, IEEE (October–December, 1999).
5. W. Kuo, T. Kim, An Overview of Manufacturing Yield and Reliability Modeling for Semiconductor Products, *Proceedings of the IEEE*, **87**(8), 1329–1344 (1999).
6. A. W. Righter, C. F. Hawkins, J. M. Soden, P. Maxwell, *CMOS IC Reliability Indicators and Burn-In Economics*, International Test Conference, IEEE (1988).
7. J. Jordan, M. Pecht, J. Fink, How Burn-In Can Reduce Quality and Reliability, *The International Journal of Microcircuits and Electronic Packaging*, **20**(1), 36–40, First Quarter (1997).
8. R. Garcia, *IC Burn-In & Defect detection Study*, (September 19, 1997).
9. T. R. Henry, T. Soo, *Burn-In Elimination of a High Volume Microprocessor Using I_{DDQ}*, International Test Conference, IEEE (1996).
10. T. Furuyama, N. Kushiyama, H. Noji, M. Kataoka, T. Yoshida, S. Doi, H. Ezawa, T. Watanabe, *Wafer Burn-In (WBI) Technology for RAM's*, IEDM 93–639, IEEE (1993).
11. P. Thompson, D. R. Vanoverloop, Mechanical and Electrical Evaluation of a Bumped-Substrate Die-Level Burn-In Carrier, *Transactions On Components, Packaging and Manufacturing Technology*, Part B, **18**(2), 264–168, IEEE (1995).

12. T. Bardsley, J. Lisowski, S. Wislon, S. VanAernam, *MCM Burn-In Experience*, MCM '94 Proceedings (1994).
13. Michael Pecht and Pradeep Lall, A Physics-Of-Failure Approach to IC Burn-In, *Advances in Electronic Packaging*, ASME, pp. 917–923 (1992).
14. W. Needham, C. Prunty, E. H. Yeoh, *High Volume Microprocessor Test Escapes an Analysis of Defects our Tests are Missing*, International Test Conference, pp. 25–34 (1992).

ESS

1. S. M. Nassar, R. Barnett, *Applications and Results of Reliability and Quality Programs*, 2000 Proceedings Annual Reliability and Maintainability Symposium, IEEE (2000).
2. H. Caruso, A. Dasgupta, *A Fundamental Overview of Accelerated-Testing Analytic Models*, 1998 Proceedings Annual Reliability and Maintainability Symposium, pp. 389–393, IEEE (1998).
3. G. A. Epstein, *Tailoring ESS Strategies for Effectiveness and Efficiency*, 1998 Proceedings Annual Reliability and Maintainability Symposium, IEEE 37–42 (1998).
4. H. Caruso, *An Overview of Environmental Reliability Testing*, 1996 Proceedings Annual Reliability and Maintainability Symposium, IEEE (1996).
5. M. R. Cooper, *Statistical Methods for Stress Screen Development*, 1996 Electronic Components and Technology Conference, IEEE (1996).

8

Why Reliability Efforts Fail

After blending the reliability processes and tools into your system, you can still fail, even with the best intentions. There are other problems that you will encounter that can stifle or seriously block your effort. In Chapter 2, we discussed the barriers to implementing the reliability process. Here, we consider how poor execution or poor follow-up can cause the reliability effort to break down.

8.1 LACK OF COMMITMENT TO THE RELIABILITY PROCESS

Commitment to a task doesn't guarantee success, but the lack of commitment is certainly a guarantee of failure. Commitment to a reliability program must come from the top management. But commitment by itself still will not guarantee success. Top management must understand what it is that they're tasking their managers to do. It is a high-level understanding of the elements of the reliability process, the cost, the requirements, the time it will take to fully implement, and what to expect from the effort. The implementers of the reliability effort must truly believe that top management has resources committed to their success. Management's everyday actions, such as signing purchase orders for equipment and materials, give believability to their commitment. They must recognize that the costs to implement reliability are easily calculated; yet the short-term results of all these actions are much more difficult to measure.

The shortsighted view of commitment to reliability is to redouble efforts toward correcting product failures by focusing on field failure analysis and corrective actions. Certainly this is part of the reliability effort, but the main effort must be to reinvent the process. The commitment must be to change the process so that failures are caught and corrected before the product is ever

Improving Product Reliability: Strategies and Implementation.   Mark A. Levin and Ted T. Kalal
© 2003 John Wiley & Sons, Ltd   ISBN: 0-470-85449-9

shipped. Management must commit to developing the know-how to change the process. At first, this know-how will come from reliability engineers, specialists, and consultants. These few individuals will impart their knowledge to the rest of the workforce. After a time, the processes will be well established, understood, and in place as part of the day-to-day ongoing activities of the company.

Reinventing the process will be a team effort. A football team has top management, a head coach, assistant coaches, many support individuals, a wide array of resources, and of course, the players. The players are the ones who have to implement top management's and the coach's plays to be successful. The players must believe in the game plan (process). To be a winner, the coaches know that they have to have a strong running game, a deceptive passing game, and a versatile kicking game.

If the line coach sees weaknesses in the right side of the defensive line, he will study the plays and the players to learn their weaknesses. If the offensive coach has a quarterback who can throw the ball into an opening that was created by deceptive running back and hits his receiver perfectly, and yet the ball falls incomplete to the ground, he knows he has to improve performance. Through observations, the coach may find that the receiver is taking his eyes off the ball. As a result, the player's hands aren't ready to clasp the ball at the precise moment. One last detail, one seemingly trivial task needs to be controlled for completion of the pass. Follow-through by everyone who is part of the process is absolutely required to be successful to win the game.

The Failure Modes and Effects Analysis (FMEA) process is not unlike the pass play in a football game. It is completed by a group of people who gather to identify weaknesses in a design. In a typical FMEA, the team may identify a small resistor, which, if it were to open, would cause power supply voltage to double, thus destroying the surrounding components. As part of the FMEA process, the group readily determines that because resistors are extremely reliable this failure is an unlikely outcome. Upon further investigation, one member of the team points out that the resistor is to be located near a corner-mounting hole. He points out that when printed circuit boards are installed and removed there is a good deal of flexure of the circuit board at and near the mounting holes. Resistors placed in close proximity to significant board flexing will cause the solder connections at the resistor to flex, possibly enough, to cause a crack. This failure may occur at the first time of flexure or over time. An open resistor or open connection to the resistor will, in this case, cause power supply overvoltage and much damage. A probable outcome of this observation will be to make a recommendation to ensure that the resistor is mounted where little flexure will take place. This means that one of the team members will be assigned that task with a date for completion. The person assigned this task must be certain that the information is given accurately to the printed circuit board designers and that they understand where acceptable resistor locations might be. Then, after the printed circuit board is fabricated, this FMEA team member must verify that the resistor is in an acceptable location. This is closure. This is follow-through. This is reliability.

In football, a lack of follow-through may range from an incomplete pass, to a missed block, to running in the wrong direction. Too many of these mistakes will lead to a lost game and a lost season. Knowing what you are supposed to do and executing every detail leads to success. Follow-through to closure when done in football, or in business, will ensure reliability of the outcome.

Follow-through in Highly Accelerated Life Test (HALT), is no different. In the FMEA process, the findings are theoretical and probabilistic. In HALT the findings are real. Remember, that the failures discovered in HALT will bear a strong correlation to the failures that may be found in the field. Correcting them before shipment is the intent of the process. After the failure is encountered during HALT, the first step is to find the actual failure. Then you must investigate further to determine the root cause and the actual physics that led to the failure. At this point in the process, you are half done. You must still identify what action is needed to prevent this failure from reoccurring. It will very likely require a design change. So one of the outcomes of the HALT process is a list of recommendations driven by failure and root cause analysis that need to be implemented. And you're still not done. You must be certain that the recommended changes have been implemented and retested to ensure that the changes perform correctly. Again, it is follow-through to closure. Without complete closure, the HALT process will not yield any improvement in reliability.

No matter how many items you find that need to be corrected in a product, your reliability efforts will fail if you disregard follow-through to closure. Finding the problems is only part of the task.

8.2   INABILITY TO EMBRACE AND MITIGATE TECHNOLOGIES RISK ISSUES

To lead, the competition companies are hard-pressed to become proficient in new technologies. This can be risky. Oftentimes, new technologies haven't been time-tested. As a result, the company risks poor return for its effort if it hasn't taken steps to mitigate this risk. For example, as electronic components become more complex, the need for connectors having a very high number of connecting pins continues to increase. If a company is planning on using a new high-density, high pin count connector just for its design, hoping that it will suffice, it will, most assuredly, lead to disaster. First and foremost, this new high-tech connector must be recognized as a potential high-risk component. You need to know little about connectors to put this connector in this category. Simply because it's new is reason enough for it to be classified as a high-risk component. Later, you must investigate the connector, its physical characteristics, how it will be installed in the manufacturing process, how it will perform in the product and how it will perform in the various field locations. Only after identifying the parameters of the connector that make it a high-risk component and taking steps to mitigate the risks [i.e., Environmental Stress Screening (ESS) testing], can the connector be deemed acceptable for use in new products. Companies that overlook the risks of any part of their new product development will suffer from low reliability.

Sometimes, a single component can be a product's Achilles heel. It's often caused by selecting a component that has not been used in the company before, and by not identifying how this component may cause problems. Usually, a team is formed to identify all the risk items on an assembly. There will be a range of risks. Some risks are higher than others. The team must identify tests that every risk item must successfully pass before the component can be an acceptable part of future products. Obviously, just selecting some tests is not adequate. Using internal and external resources, the team must identify the right tests and test environments to ensure success. Virtually every component in an assembly carries its own risk. Many are low-risk and can be set aside so you can spend more time on the higher risk items. Sometimes risks are weight, flammability, rapid wearout, or operating temperature range. The list is endless. Each risk must be identified and mitigated to the satisfaction of the risk mitigation team. Again, each identified risk item must be tracked to mitigation closure.

Using a connector as an example, the risk mitigation team may determine that the end user will use a connector 100 times in the 20-year life of the product, and that the connector manufacturer specifies that the connector be designed to withstand 100 insertion/removal cycles with acceptable reliability. This connector will be used on a printed circuit board. During production and testing, 20 insertions and removals will be consumed, leaving 80. If the end user clearly needs all 100 insertions in the 20-year life of the product, then this oversight could cause undesired failures at the end of the product's life. The risk mitigation team must either find a more acceptable connector or develop a means to produce and test the product, without unnecessarily consuming needed insertion counts. Defining the success requirements is absolutely necessary when qualifying new technologies. Not doing so is another way in which companies fail in implementing reliability.

8.3 CHOOSING THE WRONG PEOPLE FOR THE JOB

Many of the very best companies promote from within. They may take a design engineer and cross-train him/her as a manufacturing engineer. They may take a system architect and groom that person into marketing. Financial analysts can be trained to be program managers. When a company has an individual who has been performing very well, this person can be moved into an area that will continue to challenge the individual and benefit the company. This will keep individuals interested and will increase employee retention. This is a good idea if there is someone in the company who can train this individual into his/her new area. Without proper training however, the promoted employee will probably start slowly and may never grow to full competency. Digital electronic engineers can be trained as programmers as there is a considerable job similarity. Also, they will come up to speed more quickly if they work with other programmers. But training is the key. This is especially true for reliability engineering. If you do not have someone to train these talented people, they will have difficulty in delivering what is expected of them. If they are asked to go off on their own, they may not deliver what is really needed.

Reliability engineering is one of those areas that easily fall into this category. Companies that do not have a reliability program will often identify several of their best engineers from manufacturing, test, and design to become reliability engineers. Even though these individuals are hard working, talented, and respected by their peers, they don't have the tools to identify the reliability weaknesses and recommend process changes. This is one of the downsides of installing reliability in a company. Simply put, you'll probably have to hire somebody.

The reliability engineer must have a firm understanding of the processes and concepts needed to develop and enhance product reliability. This person must have the drive and initiative to install processes in a company even though there may be some resistance to a methodology new to everyone. This is a difficult task and it requires dedication and perseverance of the highest order. He/she must have a personality that adjusts to the personalities around him/her. This person must know that he/she has the full backing of the management. And finally, the reliability engineer must be a teacher. A smaller company can probably only afford one reliability engineer, and yet needs someone with all the skills. One engineer cannot do all the tasks, but must be able to impart the reliability knowledge to everyone. This process may take several years but, when done correctly, it will have trained other employees in the reliability process. Companies that try to install reliability on their own without outside help will probably fail.

8.4 INADEQUATE FUNDING

When a company chooses to implement reliability as part of its new product development, it must consider the up-front funding needed to be successful. Even before that reliability person is on board, the company must spend resources to identify and find reliability talent. Management must commit to this increase in salary expense as a minimum to get started. Very soon, the new reliability engineering hire will submit budgets to management with a timetable for implementation. Some of the budget items will be reliability laboratory space, tools and test equipment, test chamber costs (either internal or external), electrical and mechanical fixtures and training, to name just a few top items. The timing of the expenditures within the budget must be funded by the ongoing operations of the company. Financial planning must include provisions to meet these needs. A major portion of resources must be brought to bear on creating this reliability capability with the understanding that the return on investment will not be realized until after the product's release. Truly, this takes commitment and the understanding that returns are not immediate.

Early in the commitment phase, management will have high hopes for the results. As time progresses, management sees a lot of effort, many reports on product improvements, increasing development costs, and estimates of increased reliability. At this point, all they see is money going out and none coming back. This is the part of commitment where companies often fail. At this phase of the process, management has reports on field failures on previously developed products and financial reports as to what this is costing in terms of warranty dollars. Bookkeeper's ledgers are constantly adding up the cost of reliability, yet, no improvement in reliability is seen. Even though management initially understood that the return would not come until after new products were released, over time they are easily persuaded that this expenditure was a bad idea. Management must understand that they have to become true believers in the process. If they do not keep their commitments, they will certainly not be successful in their effort to initiate reliability.

One of management's major misconceptions is that they can measure increased product development time and cost. This new reliability process is delaying delivery to the customer. Upon first inspection, this is clearly true. But by accepting this delay, product development will go through fewer redesigns, which were causing the delays of the past. The new reliability process significantly reduces expenditures for multiple circuit board redesigns and software revisions, and that's just the beginning. This delay caused by the new reliability process happens only once. Because the new product will be much more reliable, engineers will not be required to develop corrections for field failures as they have in the past. This is like money in the bank. For the next new product, the same engineers will be able to apply much more of their time to new product development and much less time to fixing problems that exist in older products. Management must be patient and wait for the completion of the full product development cycle.

Companies that initiate reliability programs without follow-through will probably fail earlier than if they had done nothing. Put simply, if a company does not deliver high reliability products to its customers, the competition will. The marketplace will find manufacturers of high reliability products. If you don't install a reliability process, you'll probably go out of business; and do it even sooner with only a halfhearted commitment to reliability. You'll see that you'll spend what little money you have on something that will bear no fruit, because you gave up before the process could yield return. A weak commitment to this is even worse than no commitment. Making a commitment and failing to stay the course is a major reason companies fail when they try to install a reliability process.

HALT consumes a lot of hardware dollars. In terms of circuit board count, depending on the cost of the board and other resources, anywhere from three to six circuit boards are needed to perform HALT properly. In early product development, the engineering designers need the very first prototypes to learn how well their designs perform. After investigation, these circuit boards will usually undergo some revisions. At this point, the circuit board development is no different from what was done in the past. After learning what is needed from the prototype evaluation, several engineering changes are usually incorporated. At this point, the design team, very often, believe they are done. What makes matters worse is that given tight budgets, management may decide that the reliability team must HALT just the prototypes. This decision is disastrous.

Sometimes, several copies are made after the prototype fixes are in and yet they are parceled out to other product developers, such as programmers, test engineers, manufacturing engineers, and so on. The reliability engineers are not provided these upgraded circuit boards so that they can perform HALT. They must wait until these secondary developers have finished their activities. Here's where commitment to HALT funding is critical. Dollars must be set aside for the HALT process even though prior to HALT it is understood that the design is not complete. This is the development point in time where the designers feel the product is nearly complete. This is where the prototype discoveries were implemented and this also includes all the FMEA findings. The only thing that is not included in the new design are those failures that will be precipitated by stress testing of the circuit board and learning the stress levels needed for manufacturing [Highly Accelerated Stress Screens (HASS)]. The reliability process cannot tolerate this expenditure failure. Don't fall short at this critical juncture. Failure to do so will be a failure of the reliability process.

When the HALT process is complete and all the changes are implemented in the new design, all the reliability work done to this point, essentially, ensures that the best product that can be designed will be manufactured using the existing process. Your field failure data probably indicates that a significant part of your field failure causes are directly related to manufacturing errors. The commitment to manufacturing reliability includes HASS. At this point, engineering management feels that they have a very good product, and they do. If the manufacturing process is flawless, there is no need for HASS. (Accept the likelihood that this reliability screening process is needed because the manufacturing process probably is not perfect.)

Adding HASS to the production process adds cost and production time. These resources can be significantly reduced through proper planning. Commitment to HASS means early planning. Environmental chambers will need to be purchased and installed near the end of the production line. The mechanical fixtures required that support the product to be tested to the chamber is, in itself, a significant design task. Instrumentation and test software must also be developed as part of the process because the product will be stress-screened dynamically, while it is in operation. Adding the HASS process slows production; it's an added step. Without strong commitment, management may rationalize that HASS is not needed. (The authors do agree that if the production line processes are well controlled and the product design is reliable, HASS may not be necessary. This may seem contradictory at first glance but a review of the field failure data may show that the manufacturing process is in control.)

Remember, that the HASS process has more than one purpose. Besides it being a production screen, it can also be used as a field failure screen. After boards are repaired, they can be sent through HASS to ensure that they meet the screening standards of the production process. The HASS process identifies weaknesses in the process. It will also find weaknesses in the repair process. Whatever is fixed in repair, HASS screening will verify. When HASS is reduced to an audit process, because the process is in control, a properly scheduled Highly Accelerated Stress Audit (HASA) will ensure that there are no quality escapes. If your production line is adding contributors to field failures, your reliability process will not meet its original expectations. Companies that skimp on HASS and HASA may well fail in their reliability efforts.

There is a great disconnect between new product designers and field reliability data. Most engineers only know if their designs work for a relatively short time, typically, a year or so, then they move on to other design tasks. They do not know what the manufacturing and field failure Pareto breakdown is over time. Even when field failure reports are presented to them, they find it difficult to attach their design effort to the actual field failure data. In fact, most companies do not provide field failure data to design engineers. They will often have a department that specializes in fixing problems in the field. This disconnect is actually the broken link that allows inadequate designs to continue to propagate. Companies must communicate field failure information to the new product designers so they can evolve. Engineering management must provide this information to their designers [using Failure Reporting, Analysis and Corrective Action System (FRACAS)] to help spread the knowledge of what doesn't work.

This leads management to require the design staff to Design For Reliability (DFR). Most engineers believe they are already doing it. When designers do not receive feedback of field failure information, there is no reason for them to believe that their designs are not reliable. Designing for reliability is not well understood and DFR information certainly is not readily available. Engineering teams that have, over many years, developed DFR tools, do not publish the information because this know-how is hard-won. The reliability engineer must provide this information in training classes in order to make designers aware of things that can go wrong with what they believe to be good designs. Collecting the field failure information and presenting it in an understandable fashion to designers will greatly help them in eliminating faulty designs from new products.

When a design is complete, it is usually tested to ensure that it performs to specification. This may not be enough. Product design validation often does not include testing to identify design margins. This can be done on the test bench and/or as part of the HALT process. During testing, it is learned that under normal operating conditions a product will operate well. But if the product is not tested at the design margins, it may never be known that it is precariously close to falling out of specification, or even failure. Design changes need to be made to widen the margins for product reliability. Failures can often be attributed to designs that operate the product too close to a limit or margin. Investing the time and effort to learn the margins lead to higher reliability and fewer field failures. A lack of commitment to this effort will lead to reliability failures.

Designing too close to the operating margins is often a source for field failures that cannot be reproduced at the factory repair center. These are often referred to as *no trouble founds*; meaning that the customer sent the product back for service and the factory could not duplicate the field failure. This unit may well be returned to the field only to repeatedly fail and be returned for service, to the consternation of the customer. The actual environment in the field may be just outside the environment that existed when the product was bench-tested as acceptable, and no trouble was found. Products that fail in the field may work well after they are returned to the factory because the factory test environments and conditions do not represent the customer use environment and conditions. If this "no trouble found" product is returned to the customer, it may well fail over and over again, until someone decides to scrap this troublesome unit. Incidentally, the FRACAS system can capture this repeat field failure unit and offers the opportunity to focus on why it keeps failing repeatedly.

When management first embarks on improving reliability, they are usually driven by their awareness of excessive costs and high levels of customer dissatisfaction. Earlier in this book we pointed out that warranty costs could be significant. Setting a realistic reliability goal is very helpful in determining trade-offs that meet the needs of the business. Complex designs, that use redundancy to enhance reliability, add cost to the product. Yet, this initial cost may not be a long-term cost. These cost/warranty/reliability/redundancy analysis trade-offs should be performed and understood as part of the product development process.

Companies that don't know their actual warranty costs do not know how much money is being lost that could be returned through improved reliability (see Table I-1 in the preface). This dollar figure is actually the source of funds from where the reliability budget can be funded. One of the most important things to do initially (as management works toward their commitment toward reliability) is to put in place a warranty metric that can be tracked as the reliability process develops. Clearly, the initial reliability development funding must come from sources other than the lost warranty dollars. These dollars are not returned until after the reliability improvements have been installed. But, as reliability improvements are made, this metric will indicate how much money is no longer being lost in terms of warranty dollars. The warranty dollar measurement is a strong indicator of reliability program success. Without using this metric, a company may fail because it may well have installed reliability practices that are yielding little or no results.

There is a logical place for reliability activities. The reliability budget and estimates should be made early in product development. Design FMEAs should be scheduled near or at the end of product development but usually before any production analysis review. After corrections and improvements have been made to prototypes, the HALT process should begin. These are just a few of some of the major steps in a well-defined process. Making sure that all the reliability steps are included and in their proper order establishes a well-defined reliability process. Overdoing or underdoing reliability by not having a well-defined process will lead to failure.

Part of a well-defined process is establishing reliability estimates and budgets. The reliability budget is a breakdown of the several parts of a product in reliability terms. The end result of the several budgets is the final budget that must be met. The reliability estimates, however, comes from an analysis of the reliability of similar subassemblies already being produced by the manufacturer. Typically, whenever an assembly consists of parts and materials, very much like the parts and materials of another assembly that has been in the field, the new product will have reliability estimates similar to what has been experienced in the field. These estimates are based on experience, local to that specific manufacturer. In-house data is the best information for preparing these estimates. Reliability predictions, however, are another matter entirely.

Reliability predictions are made using dated practices and military standards that have long since become impracticable. Many people still use them, and some purchase requirements specify that these standards be used to make reliability predictions an additional requirement of the purchase specification. The authors believe that reliability estimates do little to improve a product's reliability. Reliability estimates are only as good as the data and judgment used to derive the estimates. Manufacturing and field failure data (FRACAS) that is accurately collected in your own business and in your own industry is the most accurate information available. Estimates made (using your supplier data) is better suited for determining which components have the highest failure rate. Then you can determine if the failure rate is acceptable, if the component(s) can be designed out, or the impact reduced by using fewer components. Companies that use those obsolete standards will find that it would be better to spend on reliability activities like failure modes and effects analysis, highly accelerated life testing, and accelerated life testing. Wasting resources is never a formula for success. Also, there have been several studies that have shown that reliability estimates can vary by a factor of 0.5 and 5 times the estimated value. With such a large variation between reliability estimates and observed Mean Time Between Failures (MTBF) it is hard to see their benefit.

8.5 MIL-STD 217/TELCORDIA WHAT THEY REALLY DO AND WHY THEY DON'T WORK

When the United States Government started purchasing manufactured assemblies, among the specifications that manufacturers were required to meet was a reliability metric. This measurement essentially was a measure of the time the product would function to the specification without failure. At the time this was initiated, there were no acceptable means to determine this life expectancy figure, so none had to be defined.

It was agreed that the more complicated the assembly was, the more likely it was that it would fail sooner, at least, when compared to less complicated assemblies. It was also agreed that the life expectancies of the individual components could be combined in such a way as to determine a reasonable figure for the life expectancy of the finished assembly. Systems with more components were supposed to have lower life expectancies. There was a problem, however. At that time, there was no established database of component life for all the components that went into a typical electronic assembly. So, the government went about collecting data from their sources of this information.

From its many repair facilities, that is, field repair stations, mobile repair stations, depot maintenance locations, and all over the world, the military collected failure data on component failures – how they were used, when did they fail, in what environment, and many such parameters. They had categories for benign environments as in an office; high stress environments like shipboard, tank, and helicopters. These stress factors were largely related to where the assembly was likely to be used in terms of vibration. Other stress factors were temperature, humidity, applied voltage, and more. This collection of stressors eventually grew into a uniform document now known as *Military Standard 217*. It is in the F revision as of this writing (Mil-Std-217F). It is titled the *Reliability Prediction of Electronic Equipment*.

In the title is the word "prediction," as if using a set of guidelines, set forth in tables and formulas, could be used to combine, oftentimes, many thousands of component supplier variations and individual life expectancy figures into a lump sum called a prediction. There are many problems with this method.

For the most part, the data was gathered from military personnel who were trained to repair things as rapidly as possible. Oftentimes, many components went into the assembly before it was fixed. Nonetheless, all the components were collected and these data were added to the collective database. First-line repairmen, generally, completed the repairs (the author was one of these technicians in the US Air Force, 1961–1965). These personnel were discharged in a few short years after their technical training. This means that relatively inexperienced technicians did repairs. Their mission was to get the assembly fixed as soon as possible. The number of unnecessary parts that went into the final result didn't matter at the specific location, just the speed. This tended to generate erroneous data. There were many more failed components reported than the real number of failed components.

The prediction standard went through many major revisions, from A to F. The standard was periodically updated to reflect on technology improvements but, for the most part, it was always lagging. The standard has been highly acclaimed as the foundation of reliability predictions. It began to fall into disbelief and a lack of acceptance, especially in the 1990s, because many of the consumer products that used the standard to determine a prediction were much more reliable than what the standard predicted. (Some who used the standard used multipliers from 1.4 to as high as 10 to increase the final calculated predictions because they learned, over time, that the predictions were simply very wrong.) Even so, the myth of the standard remained.

New software applications that used the data and concepts of the standard have been developed and are used by reliability consultants who specialize in reliability predictions. These too are failing to provide reasonably accurate reliability predictions. Using incorrect, unreliable information when making business and engineering decisions is not the way to be successful when implementing a reliability program.

The processes that are used to make components and that are used to manufacture complex assemblies have evolved. These changes have been occurring at a rapid pace. The standard, essentially, has fallen behind these changes, and what doesn't work is tossed aside for what does.

Reliability efforts, poorly implemented and budgeted, can easily lead to failure. Failure is guaranteed without a firm commitment to the many parts of the process. Top management must instill in the rest of the company that their commitment is sincere. This is a must in order to get the full support and belief from everyone in the firm. Caving in to other pressures midstream will also lead to the failure of the reliability process, and probably the company. Top management must commit to the new reliability program and they must stay the course if the company is to survive.

8.6 FINDING BUT NOT FIXING PROBLEMS

HALT, HASS, HASA, and FMEA will reveal problems that need attention. If not addressed, all these problems can lead to low reliability. Each issue must be carefully studied for the root cause and recommendations must be made to mitigate these problems. Each recommended corrective action must be tracked to final closure and audited by reliability engineering to verify completeness. Here, often is where the reliability effort fails.

In the rush to ship the product, the time it takes to correct a problem and make design or process changes can seriously delay delivery of the product to the customer. These delays are very visible to the bottom line and no one wants to be blamed for causing delays in shipments. Many times, the needed corrective measures are skipped, just to make shipments. This can be a reliability disaster.

Less visible are the unaddressed reliability problems that can lead to early failures in the field. The drive to ship as soon as possible, to beat competition and capture early market entry dollars can be wiped out by low reliability and poor customer satisfaction. The money gained by early market delivery can be lost due to excessive warranty claims. If the failures are serious enough to require design changes, the cost to do the design changes are considerably higher now since there are many units in the field. Fixing the problem(s) early in the development stage is the least expensive and the fastest way to make corrections. All the reliability efforts in the world will be completely wasted if the issues that need to be fixed are not addressed.

8.7   NONDYNAMIC TESTING

Product reliability testing has evolved over the years. There has been temperature testing, vibration testing, shock testing, and so on. Much of that testing is done on the product when it is operational. Nonoperational testing rarely reveals failures because the failure mode often goes away when the stress is removed. If you are going to invest the time and resources to reliability test a product, it should be done when the product is operational. Field failures occur when the system is operating, so if you want to precipitate field failures, you must operate the system under a stress test.

8.8   VIBRATION TESTING TOO DIFFICULT TO IMPLEMENT

Vibration testing is even more difficult. There is usually a mechanical apparatus or fixture that has to be designed to affix the product in test to the chamber for vibration testing. This means that the time to install the product to the vibration test fixture, run the test, and remove the product from the test fixture may seem to be prohibitive. The vibration test fixture has to mechanically couple the product to the vibration table to ensure that the forces are actually working on the product. This means that fewer units can be placed into the chamber at a time. This reduces test flow-through.

Operating the product during stress testing will require test equipment and may require test software. The test equipment adds cost. Developing the test software adds to the cost of the test process and consumes programming resources that can be used elsewhere.

Avoiding these costs will generally save money up front. The missed reliability problems will probably cost much more. Being thorough in the stress testing will return more reliability discoveries and more warranty dollars. Lack of dynamic testing is where the success of the reliability process can be lost.

8.9   LATE SOFTWARE

Software needed for the stress testing is on a critical path. If it is late, the tests cannot be done dynamically. Improperly done stress testing will result in poor reliability. Planning to ensure that the test software is ready when the stress testing is scheduled is critical for success. Poorly planned test software will be a major cause of the failure of the reliability effort.

8.10   SUPPLIER RELIABILITY

When you begin transforming your company, you may well be doing the same with your suppliers. An added function of the purchasing group is to ensure that suppliers are closing in on all the reliability issues.

Transforming the product development process to achieve higher reliability and improved customer satisfaction requires the implementation of many strategies. Doing them only half way will not lead to success.

Reference

Mil-Std 217

1. I. Knowles, *Reliability Prediction or Reliability Assessment*, IEEE (1999).

9

Supplier Management

9.1 PURCHASING INTERFACE

One of the many factors that influences the bottom line is supplier quality. The ability to receive purchased materials on time, to specification, at the quantity and quality specified, is critical to your operation. Many companies have a purchasing department, but what they really have are buyers and expediters. There is a vast difference between these two material procurement methods. Buying materials for production purposes looks easy. One picks up the phone, calls the supplier's order desk, places an order, uses a credit card, check, or purchase order, and expects on-time delivery. If your supplier has the specified material in the quantity needed, it is reasonable to expect prompt delivery. But what happens if your supplier is out of stock?

You thank your first supplier very much and call another. You continue to do this until the needed material is found. You may get the material you want in the quantity and even at the price you need. Then, you do it all over again for the next needed item. You repeat this cycle as you buy the material that goes into your product. When the bill of materials needed for production has finally been ordered, you can't be comfortable because things can still go wrong.

You need to know the following: Will all the purchased materials arrive on time for the planned production run? Will you get complete or partial shipments? Will your supplier fill the order exactly as specified? Will some other components be substituted, at the discretion of your supplier, because they were out of stock? Will the price of the purchased materials be within what is needed for you to stay within your cost margins? Chances are that these, and other problems, will occur that will negatively impact the production run, all of which will drive up your costs and lower your overall quality and reliability.

When the orders start coming in and you realize that there are discrepancies, the buyers are converted into expediters. Now, the buyer stops everything and

Improving Product Reliability: Strategies and Implementation.   Mark A. Levin and Ted T. Kalal
© 2003 John Wiley & Sons, Ltd   ISBN: 0-470-85449-9

scrambles to get the needed materials that were short-shipped. What would have been productive buying time has turned into a state of panic. Even if you are fortunate enough to get the needed material with the follow-up expedition, the mix of components that will now go into your product may cause problems that will eventually drive up costs through in-process rework, scrap, and accumulating warranty costs. All these problems would have been solved through better materials planning.

Purchased materials planning is the establishment of goals, policies, and procedures that work together to create a continuous flow of quality materials that are on time and at a price that meets the needs of your business. This is the difference between buying and purchasing. There are many variances that impact production. These variances need to be identified so that resources can be brought to bear toward minimizing their impact on your business.

An obvious planning variance would be sales volume. Is your business cyclical, seasonal, or growing at a continuous rate? Do you have some products that are declining in sales while there are others that are increasing? Do you have some products that have just completed development and which you are ramping up in production to fill anticipated orders from your marketing efforts? Do you have committed purchase orders from some of your customers with some others straddling the fence? Sales variances are major drivers in determining the need for materials. When the sales and marketing departments can generate accurate sales forecasts, the production levels can be met. From the production requirements the materials and their quantities that go into your product can be known. The longer the range and accuracy of your sales forecasts, the better will be your ability to more accurately plan materials purchases. This gives you great leverage in materials planning.

9.2   IDENTIFYING YOUR CRITICAL SUPPLIERS

Now, you have the time to identify those parts in materials that are critical, where early planning reduces risk. From your own business experience and through magazines and industry reports, you can be aware of long lead-time items. This will allow you to place orders for these critical items well in advance, before they become critical to your process. Of course, there will be a range of criticality. Many of the small components that go into your product, still require planning but usually have shorter order lead times. When business is booming for you, it is probably booming for many businesses. This usually increases the lead times for items for which your suppliers have huge demand from many of their customers. Remember, your suppliers have variances too. You may be in a business in which you need a component that is specifically designed by you and made for you by one company. The planning that is needed to ensure that this critical part will be on time is crucial to your business.

Depending on the size of your business, you will be buying materials from either manufacturers or their distributors. Establishing a good relationship with your suppliers is an absolute must. Without a doubt, it is obvious that when you receive materials that are exactly what you ordered, it is very important to make full payments on the invoice to maintain a good supplier relationship. In effect, there is a "business handshake;" you get the materials you want and your supplier gets paid on time. This is important in maintaining a good supplier relationship. Well before that, however, selecting a supplier that will satisfy your needs for the present and the future, is critical and very time-consuming.

9.3 DEVELOP A THOROUGH SUPPLIER AUDIT PROCESS

As part of your purchased materials planning, you need to know what you are looking for in a supplier before you even begin the selection process. Create a supplier audit list that identifies the important parameters you need in a supplier. These lists can be found in magazines covering the topic of purchasing, in articles and pamphlets offered by the American Society of Quality (based in Milwaukee, WI), and the knowledge about your specific business and its needs. From a combination of inputs, a supplier audit list can be constructed that will become a general template for most or all of your supplier selections.

A major part of the supplier selection is the process itself. Uniformity of the process is important when you are visiting several suppliers who may supply one type of item. Later, after auditing several suppliers, you can fairly and critically compare them against a uniform standard – your audit list. The results of the supplier audit can be used to identify strengths and weaknesses in your supplier. There is a possibility that suppliers may be unaware of their weaknesses; they might not even know that there is something they need to provide as part of their product that their customer wants. This is where you, the customer, can work toward "partnering" with your supplier.

Partnering is a concept that began in the 1980s with the Total Quality Manufacturing (TQM) boom. In its simplest form, the idea is to use the strengths of two companies to identify and improve on the weaknesses of the other. For example, the company doing the purchasing may make sporadic purchases that are difficult to fill by the supplier. The purchasing company might not view this as a problem. The supplier, on the other hand, cannot satisfy small orders and then big orders without accumulating large inventories and accepting risks that might not be in his best business interests. In this case, the supplier might be well suited to work with the purchasing company to improve their materials purchasing planning. In another case, the supplier may be delivering product that does not always meet the quality standards needed by the purchasing company. Here the purchasing company may be able to send a quality engineer to the supplier to help identify and improve their output quality. The partnering concept can be applied to virtually all segments of the business. As purchasers and suppliers work more closely together, they can minimize business risks and establish and maintain low costs with high quality.

9.4   DEVELOP RAPID NONCONFORMANCE FEEDBACK

Even with the best business relationship, on occasion, purchasers will receive nonconforming material from their suppliers. The identification of nonconforming material and the speed with which it is identified will help hold down quality costs. The sooner nonconforming material is identified in the process [Failure Reporting, Analysis and Corrective Action System (FRACAS)], the lower is the cost of recovery. Also, the time to recover from discrepant material is reduced. When nonconforming material is identified, it is to be gathered and placed in an area that is controlled so the material does not get mixed with forward production.

The discrepancy is to be identified using some form or formal process. Typically, a meeting with purchasing, manufacturing, engineering, and sometimes others is held daily to discuss the discrepant material. This reviewing group is referred to as a *Material Review Board* (MRB). Often, the location where the discrepant material is held is referred to as the MRB crib. It may turn out, that the identification of the discrepant material was incorrect, and if so, the MRB can place it back into inventory for forward production.

If the material is unacceptable and inexpensive, the best disposition may be to scrap it. Here is where supplier partnering is valuable because either the supplier or the purchaser has to pay for the scrap. If there is a good partnering relationship with the supplier by the purchasing company, they may be allowed to scrap the material at the supplier's cost, as long as the supplier can review the material at a later date when visiting without driving up the cost of shipping the material back to the supplier where they may scrap it themselves. There are several other disposition that nonconforming material can have. It may be slightly nonconforming but still used "as is." It might be used with a small amount of rework that can either be expensed by the purchasing company or billed back to the supplier. Here again, partnering helps to smooth out difficulties in these situations. If there is enough time, the material can be sent back to the supplier for corrective action. This too, can be a touchy situation, if there is not a strong working relationship between the two companies.

9.5   DEVELOP A MATERIALS REVIEW BOARD (MRB)

In any event, what is most important is that the MRB process quickly identifies the unacceptable material and works to reverse the situation, very often, with the cooperation of the supplier. Here too, the supplier is very interested in identifying unacceptable material. With an early warning, they may be able to stop current manufacturing on the very same product that is unacceptable to the purchasing company, until the matter is corrected. A rapid means of identification and feedback to the supplier via the MRB process is important. There are many software applications that are currently available that help speed this information to the supplier. Very often, both parties can share the burden of cost of the software applications. The size and complexity of these software applications vary depending on size and needs of the businesses. Depending on the software application, the feedback may take the form of a document that is automatically faxed, sent over by a modem or through the Internet to the supplier.

There is much more to the supplier management role that cannot be fully addressed in this reliability text. But, one of the most important parts of the purchaser/supplier mix, is building a partnership that understands the real needs of each party and working continuously to adapt to the needs of both businesses.

Part III

Three Steps to Successful
Implementation

10

Establishing a Reliability Lab

Installing a reliability lab in the company without proper planning and an understanding of the cost considerations can be very expensive. To begin with, the total company sales dollars and the associated warranty costs dictate the magnitude of any plan. This will guide the planner as to how many personnel will be involved in the reliability process on a day-to-day, full-time basis. This salary expense is the long-term driver because the returns on investment, in terms of recovered warranty dollars, will take several years to recover. The current salary budget must be able to absorb these expenses for this time period. This is a minimum.

Then there are the other major expenses:

- Equipment costs
- Reliability lab space
- Lab benches, desks and files, and so on
- Support tools and equipment
- Test equipment
- Mechanical fixturing [between the Device Under Test (DUT) and the chamber]
- Dynamic test devices (to operate the DUT during environmental stress)
- Consumables (Power, materials, liquid nitrogen)
- Maintenance overhead.

10.1   STAFFING FOR RELIABILITY

The reliability lab will not, in and of itself, deliver all the savings to the bottom line, but it will be a substantial part of it.

Improving Product Reliability: Strategies and Implementation.   Mark A. Levin and Ted T. Kalal
© 2003 John Wiley & Sons, Ltd   ISBN: 0-470-85449-9

To start with, one person must lead the activity. This person must have either the qualifications from other experiences or be willing to transfer current career ambitions toward reliability engineering. The latter will take a lot longer to grow. It is recommended that you seek someone from outside your present staff with the experience to build your lab and install the processes that will be utilized. His/her main characteristic skill set will include the following:

- A Reliability Engineering background,
- Highly Accelerated Life Test (HALT)/Highly Accelerated Stress Screens (HASS) and Environmental Stress Screening (ESS),
- Shock and vibration testing,
- Statistical analysis,
- Failure budgeting/estimating,
- Failure Analysis,
- Conducting reliability training,
- Persuasiveness in implementing new concepts,
- A degree in engineering and/or physics.

The salary can vary depending on experience, qualifications, and area of the country, and so on. It is recommended that you contact a recruiter who is knowledgeable in this area. (A source for this information can be found at www.salary.com.)

10.2   THE RELIABILITY LAB

The remainder of this chapter will discuss what needs to be considered in establishing the reliability lab. A matrix of suggestions for the best choices for each issue, based on small, medium, and large companies follows.

The lab space is not trivial. Besides the space needed for the HALT chamber, there will be requirements for the following:

- Liquid nitrogen tanks (if a large external tank is out of budget). Remember, that there will be some full tanks, some empty tanks and perhaps one or two partially filled tanks to contend with. The nitrogen tanks are about 30″ in diameter and six feet high and weigh several hundred pounds, when full.
- Lab benches for failure analysis, repairs, and other test equipment.
- Desk space for Intranet/Internet communications, general report preparation and so on.
- Room to maneuver test rigs in and out of the HALT chamber.
- Parking space for other equipment when not in use.

- Space for the other engineers who will take part in the HALT activities.
- Chairs for everyone, some at lab bench height and some at desk height.
- Tool cabinet(s), preferably on casters.
- Storage cabinets.
- Wall space will be needed for high-power sources, that is, shop air, coolant water sources and so on.

The cost of lab benches, desks and files and so on can add up. The lab will need a place for the following:

- The HALT operator to operate the equipment and a table or desk to prepare the test profiles, write reports, design fixtures, and so on,
- A lab bench that will be needed for test failure analysis and subsequent repairs,
- Chairs, tables, file cabinets, lab coat racks and so on.

Support tools and equipment will be needed. Some necessary tools are

- general hand tools for soldering, tightening, cutting, holding, and so on;
- a hoist for large Device Under Test (DUT) units that cannot be carried by one individual;
- thermal instrumentation, that is, thermocouples, thermocouple welders, extra accelerometers (because they fail too) and so on;
- test equipment, that is, DVMs, digital thermometers, recorders, clamp-on ammeters, portable oxygen sniffers (for leaks), oscilloscope, function generators, RF generators, and whatever your special needs may be.

You will need some sort of mechanical fixturing (between the DUT and the chamber) that will mechanically hold the DUT to the chamber table for vibration testing, such as

- drill rod and cross bars with locking nuts,
- extra bolts of various lengths, nuts and flat and split ring washers to attach holding devices to the HALT table (usually standard 3/8″ thread found in the hardware store),
- mechanical hold-downs (this may need special design for your specific needs),
- towing rope or cable to hoist heavy DUTs. (Some of these custom items may take some weeks to design and to fabricate, so planning here is important.)

Without a doubt, the dynamic test devices or instrumentation can be a high-cost item. In HALT, you must operate the DUT dynamically so that you will be able to detect failures as they happen. The instrumentation that will accomplish this may be as simple as a voltmeter and oscilloscope or it can be as complicated as a special hardware assembly with special software designed solely for these tests. (Often, this special gear can be a part of the test gear planned for the HASS in the manufacturing process later.) Here, planning is paramount. Even the length of the cables that go in and out of the HALT chamber has to be considered.

10.3   FACILITY REQUIREMENTS

There are always consumables:

- Power and light. Power can be substantial when large temperature excursions are applied during HALT.
- Liquid nitrogen, sized either by the portable dewar or large external tank. This can range into thousands of dollars per month even with the smallest HALT systems. (The wider and faster the temperature extremes the more it will cost in consumables.)
- Maintenance overhead, that is, replenishment of lab supplies and so on.

10.4   LIQUID NITROGEN REQUIREMENTS

The liquid nitrogen use cost will vary depending on the volume used and delivery frequency. Fifty gallon dewars will be the lowest cost, at first, but having a larger external tank may be the best for long-term use. The placement of the tank can be surprisingly costly. Zoning codes and local community planner preferences can make a great difference. In industrial locations, a simple concrete slab may be all that is needed. This can cost typically $10,000 to $25,000 depending on the size of the tank it needs to support. Where the community requires more esthetics, this can reach $100,000, particularly, when earthquake protection is part of the slab design specification. Things like lattice panels to cover the tank, street lighting for evening service and tank filling, special jacketed piping from a driveway located port where the nitrogen truck connects the filler hose to the tank will all be necessary; the list can be extensive.

The large tank will be beneficial in that the liquid nitrogen cost can be substantially lower by buying in bulk. There will be no need for personnel to manhandle the dewars, because the external tanks can be set up with a phone line/modem to facilitate automatic refilling. This is a great time saver, especially in the manufacturing process. Consider, as part of your big picture planning, having one tank that supports both the HALT and the HASS process. This may cost more due to the added insulated piping needed, but through careful planning even these costs can be controlled. Your tank can be larger and the volume usage costs will be lower. The number of refills will be less as well, which adds to the cost reduction.

Losses occur in the piping from the tank to the HALT chamber. This can be almost eliminated by using insulated, jacketed piping. Some nitrogen tank suppliers can provide this as part of their complete cryogenic, turnkey services. They custom design the piping as part of the whole system. This is important because where the HALT chamber and the liquid nitrogen tank are placed has a cost effect on this piping. It is typically $200 per running foot. Regular piping is less expensive but the nitrogen losses will soon add up. There will be frost buildup every time the HALT chamber is used if you select regular piping. This can add to other problems. When the frozen humidity finally warms, the resultant water may cause damage and safety problems.

All liquid nitrogen tanks are not alike. Some do a better job at minimizing nitrogen losses. Check with your supplier. It is recommended that you get a tank with an insulated output flow valve. Tanks without this valve will frost up and create a frost bubble that can be one to two feet across at the valve. This means that you cannot turn the valve until the frost has melted away. If you have a failure past the valve, you may not be able to stop the flow until the tank empties. This could be a cost and a delivery refill delay problem. Insulated valves can be shut off because they do not frost up to where they cannot be operated.

Every time you start a HALT test, the chamber has room air inside that has humidity. It will freeze during subfreezing temperature excursions and later condense when the chamber is heated. This condensation may damage the DUT. It is recommended that you add a vaporizer to the nitrogen tank that converts a small portion of the liquid nitrogen to gas. This gas can be dispensed into the chamber to flush out the humid air from the chamber itself at the start of every test. For those who select dewars, dry nitrogen bottles can be used for this purpose. Remember, that this is another tank to manhandle, reorder and have available.

You will want to budget the liquid nitrogen cost based on planned usage. A typical HALT will consume from 250 to 1,000 gallons per week. This depends on the temperature cycle rate and the temperature levels. With liquid nitrogen costing from 8 to 25 ¢/L, at the time of this writing, this cost can easily reach $1,000 per week.

10.5 AIR COMPRESSOR REQUIREMENTS

These compressors can reach $20,000 for the larger HALT units. They are best placed outside of the HALT lab so that the compressor noise does not interfere with personnel. Some compressors can be very loud. Shopping around will help you discover quiet units. Typically, the best units have full power running noise levels at 62 dBA. This is just about the noise level in a relatively quiet office. This means that for the quieter compressor you will need to obtain one that is not operated at its maximum output level. A size larger than the chamber manufacturer specifies will still do the job, work more efficiently, and be quieter. If possible, a compressor can be placed on the roof, out of the way. You must be sure to have an automatic restart feature on the compressor so that you don't have to climb up to the roof to restart the unit after power failures.

Because compressors use outside air for their source, they will add water (from the humidity) and oils (from pollution) to the airline. Proper filters and drainage facilities will be needed to dispense with this water. The HALT chambers use pneumatic piston hammers to generate the table vibration. These hammers will corrode from the water if water filtering is not done. The HALT chamber manufacturers may have a filter system at the inlets of the chamber but they will rapidly become ineffective if your facility is in a humid part of the country. Having a water filtering mechanism as part of the air compressor will help make the hammers last a great deal longer. If you use the compressed air that is already in your facility, this problem may already be eliminated. However, it may not be a good idea to use existing compressed air from your facility in the HALT chamber.

If the facility's compressed air already services other manufacturing processes, like delicate pick-and-place component equipment, then the periodic on and off of the HALT chamber with its relatively high usage, may create problems with other processes in your facility. This could create problems that will be almost impossible to diagnose. It is best to have a local air compressor for the HALT and HASS process.

10.6 SELECTING A RELIABILITY LAB LOCATION

Next, where to put the lab? The closer it is to the new product development lab, the better. You want to make the necessity of the design engineers to walk over to the HALT lab as easy as possible. Do not place the lab in another building; this often tends to make engineers reluctant to make the journey and the loss is yours. What else? It is desirable to have windows so that passersby can see the HALT in process. This is especially beneficial when you have stockholders, customers, and other management heads tour the facilities.

The lab has to be large enough to contain the HALT chamber and all the other equipment. It has to be laid out so that the product, that may have to be wheeled in, can easily be placed into the chamber. Typically a 20′ × 28′ floor plan is needed. If you know that there may be a second HALT machine in the future, the floor space may have to be large enough to accommodate the later expansion.

The lab will have to have power for the chamber, typically 480 VAC, three-phase, at 200 A. This is for the larger chambers. Less amperage is required for the smaller chambers. This specification is available from the manufacturers. You may need compressed air for your product and test methods, so plan for it.

The lab can be completely up and running in ninety days from start to finish. The faster you can get approval and have the purchase orders signed off, the better. This can be faster if sign-offs and community construction permits are expedited. In some communities, the zoning and construction permits can be exasperating in the amount of time they seem to waste. Be prepared for these delays. Get the city or town inspectors and zoning people involved early. This can help you get a fast start.

You may decide to use an outside service to get your HALT testing started quickly and later phase it into your own HALT facilities. This outside service usually has a 2- to 6-week waiting period because they also need to plan their facilities relative to typical five-day HALT exercises. This service will cost typically $2,500 per week. This is not your total cost, however.

You will still need to fabricate a test device so that your DUT can be operated dynamically. This will take time and resources. Depending on how far away the leased lab is, these costs can quickly add up. If you have to travel out of town for a week, the cost of the hotel, rental car, meals, and so on will have to be considered. In any case, you could get a jump-start on the HALT process using a leased facility. There are usually several HALT chambers at these test laboratories but still you will have to contact them for availability.

10.7   SELECTING A HALT TEST CHAMBER

In choosing a HALT chamber, there are important items that have to be taken into consideration.

First, there is the cost of the HALT chamber itself; this is the largest cost item. These chambers can cost from $50,000 to $250,000 depending on the size needed and the manufacturer. Custom chambers can be even more expensive. We have found that, as a practical matter, the published costs are relatively competitive. However, a strong negotiator can reduce the chamber costs substantially, particularly if you know you will be growing and will require increasing HALT capability and will be acquiring additional chambers in the near future. Let the chamber supplier know you are planning to do HALT *and* HASS. This usually requires two similar chambers. Buying two or more chambers can help in price negotiation, even if you are not planning to purchase them at the same time.

Before you commit any dollars to capital for HALT equipment, determine the magnitude of the dollars lost to cover product warranty. A significant portion of this warranty cost could be eliminated with HALT testing. Use this potential saving to determine how much you should be investing to purchase a HALT chamber. The savings from removing product design failures before product release is often significant enough to show an ROI in less than a year

Table 10.1   Annual Sales Dollars Relative to Typical Warranty Costs

	Small ($)	Medium ($)	Large ($)
Annual sales	1,000,000 to 5,000,000	10,000,000 to 50,000,000	100,000,000 and up
Annual warranty cost	100,000 to 500,000	1,000,000 to 5,000,000	10,000,000 and up

after product release. Table 10.1 sizes your reliability budget relative to your warranty cost. (Remember that typically the warranty cost in a company that has little or no reliability processes in place usually falls in the range of 10% of the total sales dollar.)

These figures may appear extremely large at first glance but these are typical to a variety of industries. You may think to yourself that there is no longer any need to continue reading this book because these numbers do not reflect your own company's figures. Don't be too hasty. Once you focus on all the costs that are subtracted from the sales dollars due to returned goods, rework, scrap, field service, costly design engineering changes, manufacturing process changes, hand versus automated processes, outsourcing costs, supplier relationships, inventory losses, and so on, you will learn that the cost can be very large. Your list may take a special audit on your part, perhaps hiring a quality engineering consultant who specializes in identifying the cost of quality (reliability too) would objectively reveal your true warranty costs.

Once at a division of a substantial Fortune 500 company, I was trying to establish the cost per warranty repair. I came to a figure of about $3,000 per fix. The management didn't believe it. I was asked to work on this figure with the Sr. Financial Manager to get the "right" number. After a time, we agreed that, depending on the specific product, the figure ranged from $2,700 to $3,100. I also told them that the warranty cost as a percentage of the sales figure would be in the 10 to 12% range. The top manager was in disbelief. He looked over his financial data and quickly came to the realization that it was 11.5%. He signed the purchase order for the HALT machine the next day.

A lot of the warranty costs go unmeasured and are unknown to a manufacturer. These can be in areas like how many times do personnel have to call or meet with a customer to "iron out" a problem? What did that effort cost the manufacturer? In essence, any activity that your company has to do to make your customer happy with what you have already sold him can be warranty cost. Most importantly, what additional sales were lost due to poor quality and reliability? This is a difficult figure to determine, but it is very small when you have good quality and high reliability. So, before you sit back, satisfied because you believe that you don't have a large warranty cost, think again. Make the measurement and see what the numbers really are.

The HALT chamber cost will usually be determined by the physical size of your product. If you manufacture items that are the size of a VCR, then the HALT chamber can be less expensive. However, the HASS chamber may still have to be the larger-sized unit, so that you can HASS several units at one time through the manufacturing process. Your specific planning will have to decide the best mix.

10.7.1   Chamber Size

The internal chamber size is, first and foremost, important. If it's too small, you will loose the ability to HALT parts of the product or the entire product. It may be wise to get one size larger just to be sure. The tables in the chamber are about 4″ smaller than the chamber wall-to-wall dimensions in the x and y directions. This is so that the table can move in vibration in these directions. The table has a grid of 3/8″ tapped holes that will facilitate screws and threaded rods to hold the DUT securely to the table. Most chambers have internal lights or lamps that can be swiveled for maximum adaptability. These lights take up space at the top of the chamber, so be sure that the chamber height is tall enough, even with the lighting fixtures at the top of the chamber. (The height of the chamber is really smaller by the amount of space the lamps take up.)

10.7.2   Machine Overall Height

The overall size of the HALT machine is important as it has to fit into your facility (Figure 10.1). The day will come when you will have it moved from your receiving dock and into your lab. Make sure the ceilings and door sizes allow for this. After you have selected your HALT machine, there are several other areas that need consideration. Obviously, where the HALT machine is going to be placed is important, but how you will get it there is often not a simple matter.

One installation location for the HALT machine was near a laboratory on the second floor of an engineering facility. This particular HALT machine was one of the larger units. A junior engineer was assigned the task of checking the manufacturer's dimensions of the HALT machine and making sure the unit would fit into the HALT lab. The young engineer reported the unit would fit, just barely. He even checked to ensure that the HALT machine would fit into the freight elevator that would take it from the first floor to the second. The day the machine arrived everything was ready to move the unit from the receiving dock to the HALT lab. Three very strong and burly equipment movers were contracted to do the heavy lifting. Early in the morning, as planned, the truck arrived with the new HALT machine.

It was immediately apparent that a tiny issue was overlooked. The large wooden shipping crate that contained the HALT machine would not fit into the large doors on the receiving dock, so it was temporarily set down on the

Figure 10.1   Chart HALT chamber – Courtesy of Chart Industries, Inc.

parking lot. The wooden crate material was removed. Then a rented, industrial forklift was used to place the machine back on the receiving dock. Two, smaller lifts were used to move the chamber to the lab. But it didn't fit onto the elevator. It was too tall, by two inches. The manufacturer's drawings were in error. Nonetheless, the three movers were undaunted.

With small blocks and pallet jacks the movers raised the unit high enough to remove the four metal legs from the machine, thus giving back four inches, just enough. They lowered the machine on five, one-inch diameter electrical conduit pipes, and rolled the machine onto the elevator. They reversed the process getting it off the elevator. They checked to make sure that the machine would still fit into the lab, and when satisfied, reattached the legs and placed the unit in the lab.

The lab manager had earlier hired a professional engineering firm to certify that the building was strong enough to support this five thousand pound machine so that the floor didn't cave in on office workers below. As an extra precaution, he placed a three-quarter-inch-thick, aluminum plate under the HALT machine to help distribute the weight in the lab. This precaution is usually not necessary on first floor installations, when a unit is placed on a concrete foundation. Some floors are actually a grid work raised above the foundation floor to provide room for cables and wiring and so on. It would still be wise to check the floor strength, if the unit will be placed on a tile floor such as this.

The ceiling height in the hallways from the dock to the HALT lab was higher than eight feet, but not much. The movers managed to nearly rip off an exit sign and a water sprinkler attached to the ceiling. The machine has two air filters mounted low on the unit. They were low enough for both to be damaged by careless forklift operators. The manufacturer of the HALT machine was very understanding and replaced the filters at no charge.

Needless to say, there are a lot of things that must be considered when installing a HALT machine into a lab facility. Take care to see that the details are considered.

10.7.3 Power Required and Consumption

The power required to operate the HALT chamber may reach 480 VAC, three-phase at 200 A. Be prepared to have your electrician ready. There will be power requirements for other needs. Standard 110 VAC, single-phase at 20 to 30 A may be needed throughout the lab for instrumentation and so on. Compressors often require three-phase power as well.

10.7.4 Acceptable Operational Noise Levels

Years ago, the noise levels from the HALT chambers were so excessive that the machines had to be placed in external buildings. Today, they have much lower noise levels, measured in dBA. Typically a 65- to 75-dBA noise level is acceptable when the machine is operating at its highest vibration levels. Ear protection for extended high-level tests may well be needed with some machines.

10.7.5 Door Swing

The larger units have two doors on each side. This makes the swing space required smaller. This can help lower the size of your HALT lab. Smaller machines have one door and often require the same or even more door swing space. Make sure you consider this.

10.7.6 Ease of Operation

The operation of the machines is essentially the same but how you operate them varies widely. Some controls are hard to understand while others are as simple as a cookbook.

10.7.7 Profile Creation, Editing, and Storage

The HALT test profile has to be developed by the HALT operator. It is essentially the period of the temperature cycles and the vibration and how the operator has chosen to mix them. Some software systems allow for easy copying and editing for desired changes. Make sure you test the machine you select "hands-on" before finalizing your choice. A little due diligence will save a lot of frustration when in the lab creating test profiles. The HASS process is used where there is a stronger need for an automatic profile capability. Usually, the HALT process is so empirical that an automatic system is impracticable. You will have to decide for yourself how important an automatic profile capability is.

10.7.8 Temperature Rates of Change

The rate at which the chamber temperature changes is a major selling point made by all chamber manufacturers. Some can reach 80 to 100 °C/min (This would be for the smaller DUTs that have a small thermal mass.) It is important to note, that Dr. Gregg Hobbs has written that he has never discovered a flaw as a function of the temperature ramp rate. So, maybe this "must have" feature is not that important. The ramp rate capability is important, however, in the manufacturing process, to speed throughput. This is where the ramp rate pays dividends.

10.7.9 Built-in Test Instrumentation

Some HALT chamber manufacturers only make HALT chambers. Others make a variety of environmental chambers and provide a wide variety of test instrumentation as well. Often, this instrumentation is integrated into the HALT software so that they are very compatible. This can be an important feature.

10.7.10 Safety

HALT chamber manufacturers provide second source oxygen sensors and alarms that will alert a user, in the event of a nitrogen spill, where the oxygen levels might become depleted. They are usually from another manufacturer who specializes in gas detection. To save some money, these can be purchased separately.

10.7.11 Time from Order to Delivery

From personal experience, the HALT machines are not purchased off the shelf. They have to be built to order and this time frame is usually 10 to 12 weeks from receipt of the purchase order. Some will make special provisions if you provide a letter of intent from your company.

10.7.12   Warranty

Every manufacturer has some warranty. The typical standard is two years for parts and labor. Some offer 2 to 3 preventative maintenance visits, at their costs, to ensure that your machine is operating to specification. During these visits, they may discover that you will need other repairs or offer software updates. You will probably have to bear some of these costs, if they discover problems that are out of warranty. Other manufacturers offer 90-day service and 1-year parts where you will have to cover their logistics as well. The ranges vary widely, so review the warranty policies of chamber suppliers carefully.

10.7.13   Technical/Service Support

Technical/service support is important. This may mean that there is a person who you can contact for help. It may mean that they have field service personnel who can rush to your facility to get you operational after a system failure. Sometimes, it means that they will expedite a part to you so that you can make

Table 10.2   HALT Facility Decision Guide

		Company size		
		Small	Medium	Large
HALT machine	Rent/lease	X	X	
	Buy		?	X
HALT machine operator	Rent	X	?	
	Hire operator		?	X
Nitrogen tank	Dewars	X		
	Storage tank		X	X
Concrete tank mounting slab			X	X
Multiple dewar manifold		X		
Safety mats		N/A	X	X
Training	External	X	X	X
	Internal		X	X
Mechanical fixtures		M	M	M
Test instrumentation		?	?	?
Turnkey solution		N/A	?	X
Room exhaust fans		N/A	N/A	X
Room oxygen monitors		N/A	N/A	X
Lab facilities		N/A	N/A	X
Travel costs		X		

X = likely best choice.
? = depends on circumstances.
R = rent.
M = make.

Table 10.3 HALT Machine Decision Matrix

Attributes			Company & model #						
Item #	Quantitative Model #	Example Model X1	Base #	Weighted	Company A Model - Base #	Company A Weighted	Company B Model - Base #	Company B Weighted	Sorted by weighting factors
1	Vibration level max (in Grms)	50	3	15					
2	Rate of temp change (in °C/min)	60	3	15					
3	Factory compatibility	Can use in factory	3	15					
4	Built in test instrumentation	Yes (not designed in)	1	15					
5	Machine reliability	Needs repairs every 2 months	3	5					
6	Noise level, in dBA	75	3	15					
7	Safety	Door has safety lock	3	15					
8	Machine size H"	102	3	9					
9	Machine size W"	54	3	9					
10	Machine size L"	52	5	15					
11	Chamber size H"	52	3	9					
12	Chamber size W"	52	3	9					
13	Chamber size L"	50	3	9					
14	Table size L"	42	3	9					
15	Table size W"	42	5	15					
16	Delivery in weeks	5 weeks	5	15					
17	Warranty (years)	2 year P & L	3	9					
18	Cost (base price)	$145,000							
	instrument, etc. options	$25,000	3	9					
19	Compressed air req.	120 SCFM @ 90 PSIG	3	9					
20	Max static load	800	3	9					
21	Number of nitrogen gas ducts	2	3	9					
22	Accessibility (table height to floor)	20	3	3					
23	Cable ports	4	3	9					
24	Window size (L" × W")	18 × 18, Qty 2	3	3					
25	Weight (in lbs.)	6,500	3	3					
26	Multiple dewar hook-ups	3 total	3	9					

27	Safety pads available	1 per door	5	15
28				
29				
30				
	Qualitative			
1	Software	Easy to use	1	5
2	Technical support	Yes, local	5	25
3	Graphical user interface (GUI)	8	3	9
4	Total of all HALT machines produced	125	5	15
5	Years doing HALT chambers	3	5	15
6	Years doing other chambers	None	3	9
7	Customization	Yes	3	9
8	Ergonomics	7	3	9
9				
10				
11				
12				
	Scores =		115	377

repairs yourself. Make sure you understand what the service package really is. It is important to diligently call other users of different manufacturer's chambers and see what they have to report on their support experiences. We have found that in some cases, especially when the manufacturer is not local to the user, the user soon becomes expert in the repair and maintenance of their chamber.

10.7.14   Compressed Air Requirements

As described earlier, be sure that your in-house air system can accommodate the HALT chamber needs. If not, be safe and install your own dedicated air compressor.

10.7.15   Lighting

Lab lighting is important. Make sure there is adequate lighting. It can be dark inside the HALT chamber, unless there is chamber lighting. Be sure it is part of the chamber package.

10.7.16   Customization

You may have special needs. If the chamber manufacturer will make custom machines, this can be a great asset. It may mean that you could buy a temperature only machine and have the vibration section added later. Some customers feel they only need the temperature. The authors believe strongly that temperature and vibration are the two stresses that are required at a minimum to precipitate failures. Your HALT team provides all the other stresses, voltage margining, time margining, and so on. Temperature alone is not enough.

A matrix is provided so that the reader can best decide what to do on the basis of the recommendations in the matrix (Table 10.2).

A selection matrix is provided, so that you can identify those items you deem critical in your selection of machine and manufacturer (Table 10.3).

11

Hiring and Staffing
the Right People

11.1 STAFFING FOR RELIABILITY

The reliability lab will not, in and of itself, deliver all the savings to the bottom line, but it will be a substantial part of it.

First of all, you will need one person to lead the activity. This person must have either the qualifications from other experiences or be willing to transfer current career ambitions toward reliability engineering. The latter will take a lot longer to grow. It is recommended that you seek someone from outside your present staff with the experience to build your lab and install the processes that will be utilized. His/her main characteristic skill set listed in Table 11.1 will include the following:

- A reliability engineering background
- Highly Accelerated Life Test (HALT)/Highly Accelerated Stress Screens (HASS) and Environmental Stress Screening (ESS)
- Shock and vibration testing
- Statistical analysis
- Failure budgeting/estimating
- Failure analysis
- Conducting reliability training
- Persuasiveness in implementing new concepts
- A degree in engineering and/or physics.

A reliability engineering background Look for a person with a reliability engineering experience, that is, with 5 to 10 or more years installing reliability

Improving Product Reliability: Strategies and Implementation. Mark A. Levin and Ted T. Kalal
© 2003 John Wiley & Sons, Ltd ISBN: 0-470-85449-9

Table 11.1 Reliability Skill Set for Various Positions

	Reliability skills	Consultant or reliability manager	Reliability engineer	Reliability technician
1	HALT/HASS	A	B	B
2	HALT chamber experience	A	C	C
3	HALT chamber installation	B	C	C
4	ESS	A	C	C
5	Shock & vibration	A	C	C
6	Chamber experience	B	C	C
7	Hired outside test facilities	A	B	C
8	Failure analysis	A	A	C
9	Statistics skills	A	C	C
10	Reliability budgeting	A	C	C
11	Reliability estimating	A	C	C
12	Training experience	A	A	B
13	Mentoring	A	B	C
14	Held seminars	B	C	C
15	FMEA	A	C	C
16	Has done FMEAs	A	C	C
17	Has trained others in FMEA	A	C	C
18	Component derating	A	B	C
19	Persuasive	A	C	C
20	High energy	A	B	B
21	Can show success examples	A	B	B
22	Engineering degree	A	A	C
23	Electronics (EE)	B	A	C
24	Mechanics (ME)	B	A	C
25	Physics	B	A	C
26	Business	C	C	C
27	Advanced degree	B	B	C
28	Electronics (EE)	B/C	B	C
29	Mechanics (ME)	B/C	B	C
30	Physics	C	B	C
31	Business (MBA)	A	C	C
32	Associate degree	C	C	A
33	Electronics	C	C	A
34	Mechanics	C	C	B
35	Drafting	C	C	C
36	Continued studies	B	B	B
37	Classes	B	B	B
38	Seminars	B	B	B
39	Publications	B	C	C
40	Books	B	C	C
41	Magazines & journals	B	C	C
42	Well respected by peers	A	A	A
43	Very believable	A	A	A
KEY	A = must have			
	B = nice to have			
	C = least important to have			

tools such as Highly Accelerated Life Test (HALT), Failure Modes and Effects Analysis (FMEA), component derating guidelines and so on. Analyze if he has trained others and if so how many have been trained, over what time period and if this mentoring developed other reliability engineers.

HALT/HASS and environmental stress screening (ESS) Has this candidate used or operated, or better yet, installed and used stress test chambers (HALT/ESS)? There are many things to consider, which are covered in this chapter. Can the candidate produce past stress test reports without compromising any confidentiality agreements? Is there evidence of reliability improvements, and how much and over what time period? Was the testing done in or out of house? Learn what this candidate actually did to organize the team of engineers who were involved in the testing effort. Had this skill been passed to others?

Shock and vibration testing Some stress testing is done to ensure that the product can be successfully shipped. Usually, the need for this testing is not continuous but is done on an on-and-off basis. This usually means that the test is done at a test house, where a variety of environmental and shipping tests are available on an as-needed basis. See if the candidate has this experience. Discover what was learned and what was done when design shortfalls were revealed by these tests. Has the candidate participated on shipping packaging design teams? Skill in this area can be very valuable.

Statistical analysis Design engineers have a great deal of training in mathematics but unfortunately the area of statistics is usually not part of their tool set. See if the candidate has had formal statistics training, two or more college-level semesters is good. A statistical tool that has gained great acceptance is Weibull Analysis. This is used to help identify field failure patterns. Of course, degreed engineers can learn to use this tool, the math is relatively straightforward, but having skills in statistics helps to get them going much quicker. See if they have experience with Pareto Analysis. This will help them to quickly take action to correct the most important things. Statistics can be used to demonstrate the reliability of the final product. This is useful for the management to ensure that the reliability goals have been met.

Failure budgeting/estimating Not knowing the reliability of a new product until after it is produced can be a financial disaster for a company. The ability to budget the several segments or assemblies that make up a system will help engineers to identify and focus on those parts of the assemblies that will probably be the weakest link in the system. See if the reliability engineer candidate has this in his/her background. Reliability budgeting allows early identification of high reliability risks. This can offer opportunities to make alternative choices in a design solution before it is too late for change because the product "has to be shipped." Comparing reliability estimating (that is derived from reliability estimation tools or in-house data) with budgeting closes the loop in that it is a sanity check of the budgets. Anyone can determine (guess) reliability budgets. But how well do these figures align with actual data? Look for examples of reliability budgeting and estimating. Learn how the candidate does it.

Failure analysis   Whenever anything fails, there is a reason, or in engineering terms, a "root cause". When an engineer is investigating a failure, it is common to "jump to the cause". More often than not, the cause found this way is incorrect. See if the candidate has failure analysis skills; ask for examples; ask how long it took. If the candidate has solved problems, very often, a short-term solution is implemented to fix the problem right away and then a long-term solution is added to the system as a design change. See if the candidate has been involved in these activities. Evaluate how effective and timely the corrective actions were.

Conducting reliability training   You are seeking someone who can be the seed that grows a reliability capability in your company. Discover what this person has done to pass on this knowledge to others. Have they mentored individuals, how often and how many? Also learn if there is a continuing effort by the candidate to learn more so that they stay in touch with progressing reliability technology. Have the reliability tasks been taken over by these newly qualified individuals? Has the candidate performed formal training classes, seminars, given papers, and so on? If so, what were the topics? Have they been published in journals, magazines or books? Are they aware of new software tools that will make their job easier and more efficient?

Persuasive in implementing new concepts   The authors consider the ability to recommend and persuade others to use unfamiliar tools as the "secret ingredient" to success. The evolution from a little or no reliability capability to a strong reliability capability will take several years. This takes a reliability engineer who possesses great persistence and an equal amount of support from top management. Some consider the message from the reliability group a "broken record" and it is, to some degree. The message has to be delivered on a continuous basis and the messenger has to be able to do it in a way that is persuasive, yet won't upset the rest of the staff. To sell anything, the person delivering the message has to be liked or the effort will be unfruitful. (Few of us have ever bought anything from a salesperson that we didn't like.) Study the candidate; see if this is part of their personal makeup.

A degree in engineering and/or physics   Reliability engineers interface with engineers of all kinds. No one reliability engineer can possess the skill of all disciplines, yet the person must be believed and respected by the engineering staff, as well as management. If they hold an engineering or physics degree, they will at least have a sound set of tools by which they can communicate with other engineers. See that the candidate has this tool. There are reliability engineers who have several degrees, and many have advanced degrees. Some, however, have technical backgrounds and hold other nontechnical degrees. This can be good too. *It's what they know that counts*; but having a degree is a good base on which to begin.

Some reliability engineers are members of or attend meetings of various reliability-engineering societies, several of which are the following:

- IEEE Reliability Society (http://www.ewh.ieee.org/soc/rs/) with groups all over the world.
- The Reliability and Maintainability Symposium (http://www.rams.org/) where many reliability engineers and others meet to discuss the subject, present papers, and where the American Society for Quality offers the Certified Reliability Engineering Exam.
- The Society of Reliability Engineers (SRE) (http://sre.org) and others.

Using these recommendations you have selected a reliability expert; what is the next staffing step? If one reliability person fills the bill, then you're done. Now the task is for this new person to pass on the reliability knowledge to others. If it is clear that the new reliability engineer will need support, from where can this person be found?

We recommend that you select a well-respected engineer from the existing staff. One who has a proven track record of successful designs, who has run high productivity production lines, or has a quality engineering background. Have the new reliability engineer take a strong role in selection of other internal engineers and technicians from your staff. This will help ensure compatibility and loyalty to the lead reliability engineer. Give the added staff engineer time and tools to come up to speed. If you are considering hiring a temporary consultant or a contract reliability engineer, looking into telephone directories can surface many candidates through temporary employment agencies. Often, these individuals have a wide range of experience and can get things started quickly and most effectively. If this is a permanent position, then a reliability engineering recruiter may well serve your needs.

Salaries can vary considerably, depending on experience, qualifications, area of the country and so on. A source for current salary information can be found at www.salary.com; there is a fee for the service. Five levels of reliability engineering salaries can be found there.

11.2 CHOOSING THE WRONG PEOPLE FOR THE JOB

Many of the very best companies hire from within. They may take a design engineer and cross-train him/her as a manufacturing engineer. They may take a system architect and groom that person into marketing. Financial analysts can be trained to be program managers. When a company has an individual who has been performing very well, it is often a good idea to move this person into a new department. Sometimes, but not necessarily, this will keep this individual interested and will increase employee retention in the company. This is a good idea if there is someone in the company who can train this individual into their new area. Perhaps, external training will meet the needs of the company. Without training, however, the promoted employee will get a slow start and may never grow to full competency. Digital electronic engineers can be trained to be programmers. There is a lot of job similarity. They will come up to speed more quickly if they work with other programmers. If you do not have someone to train these talented people, they will have difficulty in delivering what is expected of them. If they are asked to go off on their own, they may not deliver what is really needed. This is especially true of reliability engineering.

Reliability engineering is one of those areas that easily fall into this dilemma. Companies that do not have a reliability program will often identify some of their best engineers from manufacturing, test, and design to become reliability engineers. Even though these individuals are hard working, talented, and respected by their peers, they don't have the tools to identify the reliability weaknesses and recommend process changes. This is one of the downsides of installing reliability in a company. Simply put, you'll probably have to hire somebody.

The reliability engineer must have a firm understanding of the processes and concepts needed to develop and enhance product reliability. This person must have the drive and initiative to install processes in a company where there will be a natural resistance to a methodology new to everyone. This is a difficult task and it requires dedication and perseverance of the highest order. This person must have a personality that adjusts to the personalities around him/her. This person must know they have the backing of management all the way to the top. And finally, the reliability engineer must be a teacher. The smaller company can only afford one reliability engineer but needs all the processes. One engineer cannot do all the tasks. This engineer must be able to impart the reliability knowledge to everyone. This process may take several years but when done correctly, it will have trained other employees in the reliability process. Companies that try to install reliability on their own without outside help will probably fail.

12

Implementing the Reliability Process

Consumer demand for more reliable products will change the way businesses operate in the future. We saw this happen in the 1970s when US consumers demanded better quality autos. Consumer discontent was expressed by an increase in Japanese auto sales at the expense of the big three US auto manufacturers. When the American auto industry realized that its market share was decreasing due to inferior quality, it slowly began implementing quality programs. Change was a matter of survival. For the next several decades there was a continuous evolution of new quality programs many of which were short lived. Today, quality is a significant part of most businesses. In fact, it is widely accepted that "Quality is Everyone's Job." Experience has shown that it takes many years to fully implement an effective quality program. The same can be said about implementing an effective reliability program. Plan on it taking several years to reach full implementation and effectiveness.

12.1 RELIABILITY IS EVERYONE'S JOB

The similarities between the need for improved product quality brought on in the 1970s and the need for improved product reliability 30 years later is undeniable. The challenge is in how fast the organization can transform into designing and producing more reliable products? Business success will be based on the ability of the organization to transform into taking a shared ownership for product reliability, "Reliability is Everyone's Job."

The companies that have been most successful in achieving a reputation for highly reliable products did so by having everyone participate in the reliability process. If you establish a reliability program that fails to transform the organization into taking a shared responsibility for product reliability, the

Improving Product Reliability: Strategies and Implementation. Mark A. Levin and Ted T. Kalal
© 2003 John Wiley & Sons, Ltd   ISBN: 0-470-85449-9

results will be marginal at best. The added cost to transform the organization is small, but the cost of not doing so is large.

The reliability process can be applied to any organization and implemented at any time in a product development cycle. Implementing the process late in the development program can delay a product's release date because the process exposes design weaknesses that will likely require a redesign to fix the problems uncovered. These changes probably would have been uncovered without the reliability process when customers complained about them. It may seem to be a difficult decision to implement the reliability process late in a program because of its impact on time to market and profitability, but those profits can quickly disintegrate into significant losses from product recalls, liability suits, and high warranty costs.

12.2   FORMALIZING THE RELIABILITY PROCESS

An integral part of every reliability program is the plan detailing the activities that will take place in order to ensure success. The reliability plan must be defined and agreed upon prior to implementation. The reliability process is formalized into a document that outlines the activities for each phase of the product life cycle. A documented reliability process is a crucial ingredient for success.

The reliability plan describes the reliability activities and defines the expected deliverables for each phase of the product life cycle. After the reliability plan is developed and formalized into a document, the next step is to create awareness within the organization for this new approach to achieve improved product reliability. The entire organization must understand the new process and know what their involvement is and any budgetary cost and schedule impacts.

The reliability activities must be incorporated into the product development schedule with adequate time allotted for each reliability activity. Schedule the reliability activities at the beginning of the program, so that there are no surprises about the requirements, resources required, and impact on product delivery. At the completion of each phase of the product life cycle, review the reliability process to identify ways to improve its effectiveness for future programs. The reliability process should be continuously improved through feedback from the participants, by periodically reviewing best practices and implementing new tools/techniques to simplify and streamline the process.

In Part 4 of the book, we will present a detailed process identifying the reliability activities that take place in each phase of the product life cycle. The reliability plan has been proven to be successful and includes the necessary reliability activities for a company to design and manufacture reliable products. To be successful, you must be able to transform the organization into "doing the right things" in order to achieve product reliability. The reliability plan detailed in Part 4 represents "the right things" that need to be done in order to achieve product reliability. The way you implement these reliability activities may be different, based on your type of business and business environment. Because not all companies are alike and corporate cultures vary, the way the process is implemented will vary as well. If you implement a process of continuous improvement, the process can be tailored to best fit your business needs. By doing this, you will not only be "doing the right things" but you will also "do the right things well."

Any company implementing a reliability program or wanting to improve its product reliability program can use the reliability process presented here. These activities represent the minimum steps necessary for product reliability. By choosing only those activities that can be easily implemented, you will be making sacrifices in the reliability of the product. A common complaint made during the introduction of the reliability process by the design team is how this will delay the product launch date and increase the cost of the product. Use this as an opportunity to remind critics of mistakes made in past products and how those mistakes have impacted profitability, design resources, and product launch dates. If your products are taking longer to develop than planned, the reliability process is one tool that will help. The reliability process reduces product development time by identifying significant reliability problems early in the development cycle where they are easier and cheaper to fix.

12.3   IMPLEMENTING THE RELIABILITY PROCESS

The reliability process can be applied at any stage of the product development cycle. Ideally, the process should begin at phase one of the product development cycle. Don't wait for the next new design cycle to begin the process. There is no better time than the present to start a reliability program. The reliability process can be initiated at any stage of the product life cycle. The greatest return on investment will always be with a reliability program that is implemented at the concept phase. The goal should be to identify and fix all reliability issues as early as possible, because the cost to fix a reliability problem increases an order of magnitude in each subsequent phase. Taking a proactive approach to identify the reliability issues, early in product development, will result in a better product, with lower development costs, a shorter development time, and a greater return on investment. Often, the reliability improvements made in the development phase result in a reduction in the number of product respins later.

12.4   ROLLING OUT THE RELIABILITY PROCESS

There are many reliability activities that can be performed to improve product reliability. Some will produce more benefit than others. A list of references that will provide greater insight about these activities can be found at the end of this chapter in the bibliography.

In Chapter 7, we identified the reliability activities that provide the greatest benefit. How these reliability activities fall into the product life cycle is shown in Table 12.1.

How many of these reliability activities is your organization doing? Your present level of reliability involvement is one of the factors that will determine how best to implement the process. (Do you see anything in Table 12.1 that you are doing now?) Other factors (i.e., staffing constraints, organizational size, capital constraints, level of top down management support, product life cycle phase and time-to-market constraints) are important to consider when developing your implementation plan. *The most important factor in the implementation of the reliability process is early success.* Expect constant resistance by an overwhelming number of highly intelligent individuals who can explain in painstaking detail why the process will not work in their application. If after implementing the process you have not at least changed to some small degree the way these skeptics view the reliability activities, the whole process will suffer an early death.

There will be glitches along the way with the implementation process, especially if this is new to the organization. The reliability process is a cradle-to-grave approach. It uses continuous improvement to fine-tune the process for the organizational culture and business environment.

In order to ensure success, roll out the reliability process strategically. It is more important to achieve early, recognized successes from rolling out only parts of the process than to push the organization through the entire process. Doing too many new things at one time is an almost impossible task. In other words, it is more important to do the right things well, even if it means doing less, than it is to do all the right things to a lesser degree. Of the list of reliability steps, pick one or two and work hard to install them properly. Success in a few areas will help dissuade the skeptics and gain support. A poorly rolled out process will give added fuel to the skeptics who are trying to convince everyone that the process doesn't work. Letting them see that they may have been wrong, just a little, is more persuasive than clobbering them with a longer list of new processes. Dale Carnegie teaches how to give the skeptics a chance to save face. So do it a little at a time. They will come around and become your staunch supporters, if you let them. Some will even have the "aha!" experience.

> An "aha!" example might be, "Why do they have a gooseneck bend in the pipe under the kitchen sink; wouldn't the water go down more easily if the pipe were straight?" The answer is, "Yes; the water would go down easier but the sewer gasses would come up just as easily too." "aha!" The water in the gooseneck acts as a plug to keep the sewer gases where they belong.

Table 12.1   Reliability Activities for Each Phase of the Product Life Cycle

| | Concept phase | Design phase | | Product Phase | End-of-life Phase |
	Design concept	Production design	Validate design		
Product concept					
Reliability organizational structure	Reliability plan				
Reliability goal	Reliability budgets Accelerated life testing plans DFx	Reliability estimate Design FMEA Accelerated life testing ESS DFx	Reliability growth Design FMEA HALT POS DFx • Set up FRACAS • Reliability growth	Reliability growth Process FMEA HASS HASA DFx • FRACAS/Design issue tracking • Reliability growth • SPC • 6-sigma	HASA DFx • FRACAS/Design issue tracking • SPC • 6-sigma
Risk issues identified/mitigation plan • Continuous improvement • Quality teams • Lessons learned	Risk issues status/mitigation plan revised • Continuous improvement • Quality teams • Lessons learned	Risk issues status/mitigation plan revised • Continuous improvement • Quality teams • Lessons learned	Risk issues closed • Continuous improvement • Quality teams • Lessons learned	• Continuous improvement • Quality teams • Lessons learned	• Continuous improvement • Quality teams • Lessons learned

Note: FMEA: Failure Modes And Effects Analysis; HALT: Highly Accelerated Life Test;
HASS: Highly Accelerated Stress Screens;
HASA: Highly Accelerated Stress Audit;
ESS: Environmental Stress Screening;
POS: Proof Of Screen;
FRACAS: Failure Reporting, Analysis and Corrective Action System.

The "aha!" experience usually happens when a colleague realizes something about the design he did not know. This can happen through any of the new reliability processes, FMEA, for instance. When the FMEA process reveals an overlooked design element the lead designer usually is surprised and experiences an "aha!" all on his own. Because the revelation comes from the new FMEA process and not an individual, it, the FMEA process, is more readily accepted. With one or two more "findings from the process" this designer will be won over. Designers take pride in their work and want their design to be successful. Because they truly want to do what is right, when they have this "aha!" experience they will become strong advocates of the process. If you were to seek early support, which individual would be the best one to select?

Junior engineers will be easier to convince because this is all new to them. However, they will not be able to persuade their more experienced counterparts easily. Senior engineers have much more experience and may even have bad experiences at companies where there were reliability improvement task teams that failed. If you focus on the experienced contributors, the rest will follow. The reverse is nearly impossible. Focus on the senior level designers who are skeptical and harder to convince. Once they understand and realize the value of the process, your job will become surprisingly easier.

So, how many of the reliability activities in Table 12.1 is your organization doing? If the answer is none to very little, then you will want to take a slower, incremental approach in implementing these new processes. One reason for this is that the reliability activities do not take place in a vacuum. You cannot just hire a group of reliability engineers and tell them to make the product reliable. This strategy will most likely fail. Remember that reliability is everyone's job. Reliability engineers do not design products, the design team does. Many of the reliability activities require the participation of the design team in order to be successful.

An FMEA requires a design team of cross-functional members participating to identify failure modes and safety issues. Team members are then tasked to remove those issues that have a high risk. A HALT test can require several months of preparatory work with the design team. They need to develop fixtures, test procedures, test software, and test access. During the HALT testing, which can last one to two weeks, engineering support will be needed to fix precipitated failures and to identify their root cause. Likewise, risk mitigation, FRACAS, and DFx [Design For Manufacturing (DFM), Design For Service (and maintainability) (DFS), Design For Test (DFT), Design For Reliability (DFR) etc.] are all activities that require the participation of other functional groups. Because of this, you can bring product development to a standstill if you implement everything at once. The famine-to-feast strategy for implementing a reliability program when all these activities are new to the organization will be devastating.

In fact, there are only a few reliability activities that can be done by reliability engineers alone. These activities would include reliability budgets and estimates, component accelerated life and environmental stress testing, reliability growth tracking, and reliability demonstration. In addition, some of these activities (reliability budgets and estimates, reliability growth tracking, reliability demonstration) contribute little or nothing to improve product reliability.

How do you decide the best strategy for implementation? Which of the reliability activities are the most important? Which must be implemented in order to achieve product reliability goals? Not surprisingly, there is no one program that will fit all businesses. If you produce a product for space travel, life support, nuclear reactors, and other mission critical applications, the reliability effort will be significant and comprehensive. If you produce extremely low cost, short product life, disposable products, then you are at the other end of the extreme. However, our experience has found that there are a few reliability activities that everyone should be doing in order to achieve the needed reliability improvements. These activities can be found in Table 12.2.

The reliability of a product is determined in the design phase of the product development. Once the product is designed, the reliability can be degraded through poor manufacturing, inadequate testing, or troublesome suppliers. Remember the old saying about "You cannot test in quality." The same can be said for reliability. You cannot manufacture or test in reliability. Product reliability is determined by how well the design meets the design specifications for the different-use environments and conditions and for the expected time of use.

The two most powerful product development reliability tools, which will improve the reliability of a product, are FMEA and HALT. If your business lacks a reliability program, then, the first reliability activities that need to be installed are FMEA and HALT. A design FMEA is a powerful tool, which uses the design team's aggregate expertise to create a synergy that results in identifying design problems often overlooked in design and design review. Of course, you can discover these same issues late in the product life cycle, but it will be at a much greater cost to the company. HALT is the best way to take a working design and precipitate the most likely reliability failures that will occur in the field. If you have no reliability program in place, then implementing these two reliability tools first into product development will be what is most needed.

Once you have successfully implemented these tools, the next step is to develop DFR guidelines. We do not know of any DFR guidelines that can be purchased, so you will likely end up developing your own. The design guidelines should be based on lessons learned, focusing on the Pareto of top reliability problems you have had in the past. If capacitors are high on the Pareto chart, then, you will want to develop a capacitor selection and use guide. The DFR guidelines should also include issues such as derating. The best way to develop DFR guidelines is to create a Pareto of past reliability issues and, starting

Table 12.2 Reliability Activities for Each Phase of the Product Life Cycle

Functional activities	Concept phase		Design phase		Production phase	End-of-life phase
	Product concept	Design concept	Product design	Validate design		
Every business needs		Design FMEA	Design FMEA	Design FMEA	Process FMEA	Continuous improvement
				FRACAS	FRACAS	FRACAS
				HALT	SPC/6-sigma	SPC/6-sigma
				Early production HASS		
Will make life easier	Reliability goals	DFR	DFR	DFR	DFR	HASA
				POS	HASA	
				Reliability growth		
Nice to have		Reliability budgets	Reliability estimates		100% HASS	
Do if required					Reliability demonstration	

Note: SPC: Statistical Process Control.

with the biggest issue, develop reliability design rules to eliminate repetition of these problems.

12.5 DEVELOPING A RELIABILITY CULTURE

Product reliability must be everyone's job. To achieve this work philosophy, you will need to transform the organization's culture into one where everyone talks about product reliability issues. Getting an organization to this point will take time. If you are just beginning to implement a reliability program, the following three processes need to be in place before the program rolls out.

1. Formalize the reliability process in a document
2. Implement top down training for the new reliability process
3. Prepare a reliability process implementation plan.

The first step is to define the reliability process that will be followed. Part 4 of the book provides the detailed reliability process for successful product reliability.

The second step is to develop training to educate the organization on the new reliability process. The training should be rolled out in a top-down approach. Senior and middle level managers need to buy into the process before it is disseminated to all other levels of the organization. If there are issues raised by senior and middle management that are not resolved before rolling out the training to the masses, you are unlikely to get the buy-in needed for success.

The final step involves developing a credible implementation plan that transforms the organization into a culture that is focused on reliability issues and able to achieve the reliability goals. The implementation plan will be different for different-sized companies. For very large companies, consider using a seven-infrastructure approach as outlined in the book "A New American TQM" by Shoji Shiba, Alan Graham, and David Walden, Productivity Press, 1993, Chapter 11. Use the seven-infrastructure approach to transform the organization into a culture that relies on the new reliability process to ensure product reliability. The organizational infrastructure approach identifies seven activities that need to take place. The seven activities are as follows:

1. Goal setting
2. Organizational setting
3. Training and education
4. Promotion
5. Diffusion of success stories
6. Incentives and awards
7. Diagnosis and monitoring.

A New American TQM provides an effective framework that can also be used to implement a reliability program in an organization. Implementing a reliability program is no different from implementing a quality program. Today, most companies have quality programs in place. Do you remember how difficult it was to implement these programs and how many of them died within 6 to 12 months? In most companies, there was a significant amount of resistance to changing the way they manufactured and developed products. It is difficult to change the way an organization operates. In essence, you are changing the culture of the organization. The changes are usually slow to take hold and often take years to fully implement. Therefore, take time to review the organization's effectiveness in past rollout programs like Total Quality Manufacturing (TQM), quality circles, and continuous improvement. Identify what worked well and what did not. This way, the organization can learn and benefit from this experience. Then, define an achievable plan for implementing the reliability process.

12.6   SETTING RELIABILITY GOALS

It's time to set reliability goals. There are two types of goal setting that take place in a reliability program. First, there are the high-level nonprogram-specific goals. The highest-level goals are the mission and vision statements for the organization. The mission and vision statement addresses the business need for improved product reliability. Before you create the mission and vision statements, determine the business environment driving the need for greater reliability. What is the customer's perception about your product's reliability? Is it different from that which you have measured, observed, or perceived? What is the perceived reliability of the market leader? Is improved reliability a strategy to maintain or gain market share? Has there been a problem with highly publicized product recalls? Are product liability lawsuits a problem?

Knowing the answers to these questions can prevent the implementation of a very costly and misdirected reliability program. There are costs associated with improving product reliability. These costs affect the bottom line. When implemented effectively, they will bring significant long-term gains. However, a reliable product that is not cost competitive can have an adverse effect on market share.

If you are implementing a reliability program for the first time, there should be high-level goal setting focused on the implementation of the reliability program. These goals focus on the following:

- Forming the reliability organization
- Installing the reliability lab
- Defining and documenting the reliability process

- Implementing a reliability process into the organization
- Developing reliability design guidelines and checklists
- Implementing FMEAs
- Implementing HALT, HASS, & HASA
- Implementing FRACAS.

The second sets of goals are the low-level goals. The low-level goals are program or product specific. The goals are measurable, result oriented, customer focused, time-specific, and support the high-level goals. They can be different for different products. Examples of program goals would include the following:

- Will perform without failure for a specified time and under defined environmental use conditions
- Reduce repair time
- Reduce product development time through fewer design spins
- Reduce product development costs through fewer design changes
- Improve manufacturing first-pass yield (through improved design margins).

The goals that you define should be measurable and supportable in the business environment.

12.7 TRAINING

The greatest benefits of a reliability program are the design improvements made before the first prototype is ever built. Unfortunately, reliability usually takes a back seat in the early phases of the product development cycle. Design engineers do not like to be told how to do their job. We often assume that the people we hire are experts or at least competent in all facets of their job. Unfortunately, this is not always the case. It is not that they are bad designers; it's often simply that they lack the knowledge and skills required to improve a design for reliability. Simply providing training to designers on making the right design decisions will not improve reliability. While in the proactive phase of the program, it is important to provide training to the design team, so the decisions they make will lead to reliable product designs.

Focus the training in the areas where the reliability of the product has been a problem. For example, suppose your product has had an occasional problem with catching on fire. The product was designed with safety fuses that were supposed to prevent failures that led to fire. An investigation reveals that the fuse was improperly designed. Because the proper selection and use of fuses are extremely important to the product, it is important and beneficial to provide training to the design team on the proper use of fuses. The training should be offered periodically because new employees cycle in and out of the organization and, sometimes, the past lessons learned are forgotten. There are several ways to provide training within the organization. Some of the more common approaches are as follows:

1. Develop training internally using in-house expertise
2. Send employees to symposium classes and conferences
3. Use outside experts/consultants to teach classes
4. Encourage higher-level education
5. Provide a library of books on the subject.

Ideally, there is an individual within the organization who is an expert on the design and the use of fuses. Work with this individual to design a training class on fuses. Providing training using in-house expertise is always the preferred method when the expertise exists. This is a low-cost approach that has the added benefit of communicating where the resident expertise is for a particular subject. A large organization will typically have many experts whose knowledge is often underutilized in solving known problems. Make the training materials, presentations, and so on easily available for those who could not attend the training when offered. Be sure to identify the author of the training, as this person is the in-house resource on the subject.

While individual companies will have unique and specific training needs, the following subjects are needed universally:

1. Capacitor selection and use
2. Redundancy
3. Connector selection and use
4. Derating guidelines and use
5. Mechanical reliability
6. Torque and hardware stack-up
7. EMI/RFI shielding
8. Electrostatic Discharge (ESD) protection and susceptibility
9. Solder reliability
10. Corrosion
11. Cooling techniques
12. Materials selection.

12.8   PRODUCT LIFE CYCLE DEFINED

The best thing about a reliability program is that it can be applied successfully at any stage of the product life cycle. The reliability process can be applied to a product revision, derivative product, new platform product, or a leapfrog technology. The timing of the process does impact the level of risk taken, the level of effort required, the resources required, or the time frame necessary to ensure product reliability.

In Part 1 of the book, we described why product reliability is vital to any business that wants to compete in the twenty-first century. In Part 2, we developed the reliability tools that are needed to improve product reliability. In Part 3, the process of forming a reliability team and installing a reliability facility was developed. The only piece missing from the puzzle is the implementation of a reliability process. Therefore, in Part 4, we shall put the pieces together to form the reliability process.

The reliability process is a comprehensive cradle-to-grave approach to improve product reliability. The process should be part of a continuous improvement program that applies lessons learned from past products to continuously improve next-generation products. The product life cycle consists of six phases:

1. Product concept phase
2. Design concept phase
3. Product design phase
4. Validate design phase
5. Production phase
6. End-of-life phase.

These are shown graphically in Figure 12.1. Because each company may define the product life cycle phases differently, we briefly describe each phase so you can align them to your unique product development structure.

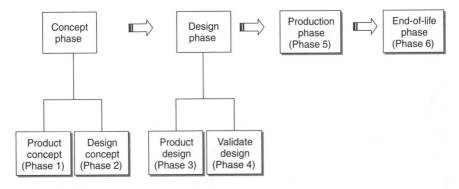

Figure 12.1   The six phases of the product life cycle

12.9 CONCEPT PHASE

In the concept phase, the product is conceptually defined sufficiently for a team to design the product. First, the product concept is defined based on market and business needs. The design concept is developed defining the product architecture, physical features, inputs & outputs, assumptions, and so on. The concept phase defines the design requirements, constraints, features, and limitations that will be used to direct the design team. The concept phase may also include a design priority selection list (i.e., in order of priority: cost, time to market, performance, reliability, and manufacturability) that designers use uniformly to make design trade-offs. A list of the desired outputs from the concept phase is as follows:

Product concept phase deliverables

- Market-driven product concept
- Product features requirements
- Product functions requirements
- Performance specifications
- Product positioning
- Market/business-driven time-to-market date
- Staffing required to achieve time to market
- Capital required to achieve time to market.

Design concept phase deliverables

- System & subsystem design architecture
- Preliminary design concept
- Design specifications
- Define what design needs to do
- Define what design will not do
- Define design decision trade-offs (i.e., in order of priority: cost, performance, time to market, size, weight, etc.)
- Maintenance and serviceability requirements.

12.10 DESIGN PHASE

The next phase of the product life cycle is called the *design phase*. It too, is composed of two separate phases. The design phase begins with the product design phase where design teams create the design details necessary to achieve the concept requirements. It is in the design phase where working prototypes are developed for design validation. The design phase is also where the product documentation package (Printed Circuit Board (PCB) design, schematics, Bill of materials, mechanical drawings, etc.) is created.

The product design phase is followed by the design validation phase where the working prototypes are tested to verify that the design meets the requirements called out in the concept phase. At the end of the design validation phase, the design is verified to be manufacturable, testable, and serviceable. By the end of the design validation phase, the product cost and profit margins are well understood along with strategies to reduce product cost. A list of the desired outputs from the design phase is as follows:

Product design phase deliverables

- Schematics
- Theory of operation
- Bill of materials
- Mechanical drawings
- Product costing
- Working prototypes (hardware, software)
- Supplier selection
- System & subsystem test strategies
- Test fixtures
- Manufacturing fixtures.

Validate design phase deliverables

- Verify design performance to specification
- Verify adequate design margin
- Verify production test & fixturing
- Verify manufacturability of product & fixturing
- Engineering change orders Implemented and changes verified
- Shippable product.

12.11 PRODUCTION PHASE

The production phase begins with the transitioning of the design for production and manufacturing. The engineering effort in the production phase has significantly reduced to a support effort. It is in the production phase that manufacturing begins ramping product to meet customer demand. The production phase activities are focused on supporting product manufacturing, test, and customer support. The activities that take place in the production phase are as follows:

Production phase deliverables

- Manufacturing process control
- Volume production tooling
- Supplier management
- Inventory control
- Cost-reduction programs
- Cycle time improvements
- Defect-reduction programs
- Field service & tech support programs.

12.12   END-OF-LIFE PHASE

The last phase of the product life cycle is called the end-of-life phase. All products have a useful life and eventually reach a point of obsolescence. The end-of-life phase includes all activities associated with the eventual termination of the product. This is the last phase of the product life cycle and the one most often overlooked.

The six phases of the product life cycle are discussed in detail in the final four chapters of this book.

End-of-life phase deliverables

- Eliminating obsolete parts and materials from inventory
- Disposing of manuals, documents, and so on no longer needed for product support
- Introduction of a transition product.

12.13   PROACTIVE AND REACTIVE RELIABILITY ACTIVITIES

The reliability activities in the product life cycle can be considered either proactive or reactive. The proactive activities consist of everything that can be done to improve product reliability and serviceability before the first customer shipment. In essence, we are trying to remove customer failures by identifying what is likely to fail in the field. The reliability activities turn from proactive to reactive once the product is released for manufacturing. Design changes made after this point are more expensive and take longer to implement. In fact, a design change after a product is released receives greater scrutiny and is less likely to be implemented because of its impact on the bottom line. The same change requests made early in the design cycle are likely to be implemented because it is significantly less expensive to implement, does not impact product in the field, and the design team has not moved on to the next project. Design changes represent business decisions because they impact product and development cost, product release date, and warranty costs. The focus clearly needs to be on optimizing a DFR in the proactive portion of the product life cycle.

The proactive region of the reliability program identifies all potential reliability issues before the products are shipped to customers (Figure 12.2). The proactive phase identifies potential risk and safety issues and resolves all potential reliability problems, which are likely to occur. The proactive reliability activities are as follows:

1. Failure Modes and Effects Analysis (FMEA) before design is complete
2. Applying lessons learned
3. Appling design guidelines:
    (a) Design For Reliability (DFR)
    (b) Design For Manufacturing (DFM)
    (c) Design For Tests (DFT)
    (d) Design For Serviceability and maintainability guidelines (DFS).
4. Identify, Communicate and Mitigate (ICM) approach to mitigate technology risk
5. Complete design simulation and modeling
6. Complete design specs and requirements before design phase
7. System and subsystem reliability budgets and estimates
8. Highly Accelerated Life Tests (HALT)
9. Four corners testing, testing at design margins.

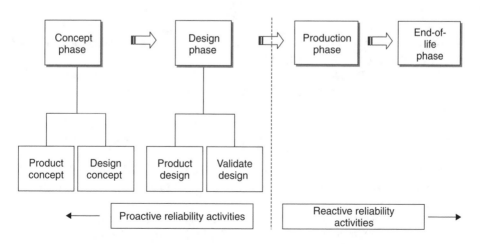

Figure 12.2   Proactive activities in the product life cycle

These reliability activities will provide the greatest benefit early in the development program. By applying these activities early in the concept and design phase, you reduce development time, NonRecurring Engineering (NRE) costs, and the number of design spins. By performing these steps early in the development cycle, the design is likely to be error free the first time.

The reliability activities become reactive once the product is approved for manufacturing. The reactive reliability tools are used after a product is released to manufacturing. It is then, we find ourselves scrambling to deal with difficult issues regarding product recalls, product alerts, and retrofits. Finding these problems is often costly and time-consuming. When these problems are found late in the development cycle, they usually lead to expensive and time-consuming design changes, which end up delaying the product release date. Some of the reliability activities used in the reactive phase are as follows:

1. Highly Accelerated Stress Screening (HASS)
2. Proof Of Screen (POS)
3. Reliability growth curve
4. Design Maturity Testing (DMT)
5. Functionally test at design margins
6. Failure Reporting Analysis and Corrective Action System (FRACAS)
7. Root cause failure analysis
8. SPC
9. 6-sigma.

References

Reliability process

1. G. Novacek, Designing for Reliability, Maintainability and Safety, *Circuit Cellar*, **January**(126) 28 (2001).
2. S. M. Nassar, R. Barnett, *Applications and Results of Reliability and Quality Programs*, 2000 Proceedings Annual Reliability and Maintainability Symposium, IEEE (2000).
3. R. Green, An Overview of the British Aerospace Airbus Ltd., *Reliability Process, Safety and Reliability Engineering*, British Aerospace Airbus Ltd., IEEE, Savoy place, London WC2R OBL, UK, (1999).
4. D. R. Hoffman, M. Roush *Risk Mitigation of Reliability-Critical Items*, 1999 Proceedings Annual Reliability and Maintainability Symposium, IEEE, pp. 283–287.
5. I. Knowles, *Reliability Prediction or Reliability Assessment*, IEEE (1999).
6. J. W. Evans, J. Y. Evanss, B. Kil Yu, Designing and Building-In Reliability in Advanced Microelectronic Assemblies and Structures, *IEEE Transactions*

on *Components, Packaging, and Manufacturing Technology*, Part A, 20(1), 38–45 (1997), IEEE (March 1997) & Fifth IPFA '95 Singapore.

7. H. Caruso, *An Overview of Environmental Reliability Testing*, 1996 Proceedings Annual Reliability and Maintainability Symposium, pp. 102–109, IEEE (1996).

8. U. Daya Perara, *Reliability of Mobile Phones*, 1995 Proceedings Annual Reliability and Maintainability Symposium, pp. 33–38, IEEE (1995).

9. S. W. Foo, W. L. Lien, M. Xie, E., van Geest, *Reliability by Design a Tool to Reduce Time-To-Market*, Engineering Management Conference, IEEE 251–256 (1995).

10. W. Gegen, *Design For Reliability – Methodology and Cost Benefits in Design and Manufacture, The Reliability of Transportation and Distribution Equipment*, pp. 29–31 (March, 1995).

11. W. A. *Golomski, Reliability & Quality in Design*, W. A. Golomski & Associates, Chicago, pp. 216–219, IEEE (1995).

12. D. J. Leech, Proof of Designed Reliability, *Engineering Management Journal*, 169–174 (1995).

13. W. F. Ellis, H. L. Kalter, C. H. Stapper, *Design for Reliability, Testability and Manufacturability of Memory Chips*, 1993 Proceedings Annual Reliability and Maintainability Symposium, pp. 311–319, IEEE (1993).

14. J. Kitchin, *Design for Reliability in the Alpha 21164 Microprocessor*, Digital Equipment Corporation.

Part IV

Reliability Process
for Product Development

13

Product Concept Phase

Product development begins with the concept phase. It consists of two parts, the product concept and design concept (discussed in Chapter 14). In the concept phase, a decision is made to develop a new product. It is in the concept phase in which marketing, engineering, operations, and filed inputs yield product concept requirements. It does not matter if the product is a new platform or a derivative product; the process is the same. The concept phase is often conducted in a vacuum between senior engineering and marketing management. The decisions made during this time have a dramatic impact on the entire organization. It is in the concept phase that the product is defined on the basis of market needs, customer focus, product features, product cost, business fit, and product architecture. This may seem like a strange time to begin activities regarding product reliability because there's so little known about the actual product itself, after all it is only a concept. No detailed design effort has started, so there's no work to be done on improving the design. The main reliability objectives in the concept phase are to form the reliability team, define the reliability process, establish product reliability requirements, and a first-pass risk assessment to Pareto previous reliability problems. These top reliability problems become design constraints for the design concept phase. A summary list of the reliability activities performed in the product concept phase along with the expected deliverables is shown in Table 13.1.

During the concept phase, design decisions are made which may require new technologies, materials, and processes. The decisions made here can impose significant risk to the design, manufacturability and rampability, testability, serviceability, and product reliability. The product concept phase represents the first opportunity to identify significant risk, which can jeopardize the success of a program. The risk issues impact the entire organization. These issues need to be identified, agreed upon, and a risk resolution strategy [Identify, Communicate, and Mitigate (ICM) plan] laid out. If the risk issues can be

Improving Product Reliability: Strategies and Implementation. Mark A. Levin and Ted T. Kalal
© 2003 John Wiley & Sons, Ltd ISBN: 0-470-85449-9

Table 13.1 Product Concept Phase Reliability Activities

Participants	Product concept phase	
	Reliability activities	Deliverables
• Marketing	1. Form reliability organization and responsibility.	1. Reliability team formed with agreement on the reliability activities.
• Design engineering	2. Define the reliability process.	2. Description of the reliability activities that will be performed.
• Reliability engineering	3. Define product reliability requirements.	3. Product level MTBF, MTTR, availability defined.
• Field support/service	4. Capture and apply external lessons learned.	4. Pareto top VOC reliability issues and recommendations.
	5. Develop risk mitigation form and have meeting to review each risk issue.	5. Completed risk mitigation form and meeting results in acceptance of risk issues and planned mitigation.

Note: MTBF: Mean Time Between Failures; MTTR: Mean Time To Repair; VOC: Voice Of Customer.

identified and agreed upon early in the program, the organization will have the needed time to plan and mitigate all significant risk before first customer ship. The product will be more reliable if the risk issues are resolved prior to product release.

There are five major reliability activities that take place in the product concept phase. The five activities are the following:

1. Establish the reliability organization
2. Define the reliability process
3. Define product reliability requirements
4. Capture and apply external lessons learned
5. Risk mitigation.

13.1 ESTABLISH THE RELIABILITY ORGANIZATION

Every reliability effort requires a staff to implement the reliability process. The reliability staffing may be small, and in many cases a single reliability engineer will suffice. There is no set rule as to how large the reliability team needs to be for success. Staffing the reliability team at 1% of the design team size is a good starting point. (The topic of staffing for reliability is covered in

Chapter 11. Chapter 8 also addresses some of the problems with selecting the wrong individuals.)

In forming the reliability team, selecting an individual who has strong leadership skills is the key. If the reliability program has been established, then selecting an individual with management or engineering skills will suffice. Small organizations may have only a single engineer supporting all the reliability activities. If you have a small organization, then select someone with good management skills.

Finally, the reliability workload is not constant in each phase of the product life cycle. The workload is small in the concept phase and reaches a peak in the design phase where accelerated life and Highly Accelerated Life Testing (HALT) consume significant resources. If you have a quality organization, then the reliability effort can transition to a quality effort in the design validation phase. Having quality and reliability teams engaged jointly during design validation ensures shared ownership, communication, and cooperation between the two functions. Having the two teams report to the same manager will ensure that there are no barriers formed between the two groups.

The reliability design members should have a strong technical background with past experience in product design. The quality team's members should have a strong technical background in manufacturing, process control, and Failure Reporting, Analysis, and Corrective Action System (FRACAS). Having separate teams is preferred because it is rare to find individuals with strong design and manufacturing experience.

13.2   DEFINE THE RELIABILITY PROCESS

The reliability process is outlined in Chapter 12 and presented in detail in Part 4 of the book. Use the detailed reliability process and tailor it to suit your particular needs. The goal should be to define the process up front, schedule it into product development, and then get everyone to agree to follow it. Once the process is defined, plan on conducting training in the concept phase to educate everyone on the process.

13.3   DEFINE THE SYSTEM RELIABILITY REQUIREMENT

An integral reliability activity in the product concept phase is setting the product reliability requirement. It should be market-driven and focused around the targeted customer requirements. It is usually described in terms of Mean Time Between Failure (MTBF), availability, serviceability, and maintainability requirements. In setting the system level reliability requirements, consider the previous product's reliability performance and how they compare with key competitors (benchmarking).

Ask yourself the following:

- Are you losing market share due to unacceptable product reliability?
- What increase in market share do you expect if the reliability of the product is increased?
- What improvement in profit margins can you expect with improved product reliability?
- What is the customer expectation for the reliability of this product?
- What is the customer willing to pay for improved product reliability?

Use the answers to better describe your reliability goals and objectives.

13.4 CAPTURE AND APPLY LESSONS LEARNED

It is in the product concept phase that we take the time to reflect on the reliability problems from past programs. It is hard to understand why companies continually repeat the same mistakes. Reliability problems continually reappear even though it is relatively easy to apply lessons learned to new designs. Large companies with divisions scattered around the globe have difficulty communicating lessons learned. The larger the company, the greater the problem. As a result, the mistakes made in past programs get repeated until they get high visibility, often through extreme customer dissatisfaction or significant financial exposure. By the time this occurs, years have passed, resulting in a logistics nightmare to fix the problem. Without a formalized process to capture, train, and apply lessons learned, past reliability problems will continually be repeated. These lessons learned should be incorporated into a Design For Reliability (DFR) guideline. A design review checklist is another effective way to verify that lessons learned are getting implemented. The solution needs to be part of a formal process to capture and implement lessons learned.

It is the external lessons learned which are most relevant in the product concept phase. (The internal lessons learned will be captured and applied in the design concept phase of the product development cycle.)

There are four areas to focus on when capturing the external lessons learned. These are as follows:

1. Conduct an external Voice Of the Customer (VOC). The customers in this case are the end users, product support groups, and individuals who service and repair the product.
2. Review the FRACAS reports, then Pareto the field failures.
3. Identify past product recalls and safety warnings.
4. Review the customer complaints file.

Conduct an external VOC regarding reliability issues on previous products. Focus the VOC around product reliability, serviceability, and maintenance issues. This can be done through in-person interviews, phone surveys, mail questionnaires, or the Internet. Some methods will be more effective than others and experimentation may be necessary to determine the best method. It may be a good idea to use a third party for this activity because internal eyes and ears may not be sensitive to every complaint. Whichever method you choose, keep the survey unbiased and the questionnaires the same for all who are surveyed. A weighting mechanism can be used to differentiate the importance of issues to the customer so a meaningful priority can be associated with each issue. Test the questionnaire in advance (internally and externally) to ensure there's no confusion regarding the language, weighting scale, and intent of each question. Once the data is collected, tabulate the results to determine what reliability issues must be included in the concept phase documentation.

Another source for capturing lessons learned is from the FRACAS database. FRACAS tracks field failures, ranks them by severity of occurrence, documents root cause, and tracks resolution effectiveness. It is important to know the root cause associated with each failure. If the root cause has not been determined, then it may be necessary to launch an activity to determine the root cause prior to completing the design phase. If the root cause cannot be determined, then there is no assurance that the problem will not resurface in the next product.

One big challenge that every organization faces is determining the root cause of a failure. One reason for this is that many organizations lack the technical skills and expertise to link complex failures with the root cause. Getting to the root cause requires an understanding of the physics behind the failure. Once you have identified the root cause of a failure, correcting the problem becomes easy. If you don't identify the root cause, the fix can end up correcting a nonexistent problem, and some time later the problem will resurface.

One technique used to determine the root cause of a failure is to ask why, five times. To illustrate this point consider a standard flashlight that does not illuminate when it is turned on. The problem is traced back to the light bulb not getting the required battery voltage. The problem is further traced to the battery voltage not making good electrical contact with the spring. At this point it is hard to determine what is the root cause of the problem. The problem could be corrosion on the spring, an incorrect or weak spring, spring not installed properly, spring material is too resistive, contamination on the battery contacts, or an intermittent battery.

So we ask the question a second time, "Why is there poor contact between the battery and the spring?" The answer to this question turns out to be a loose spring. There are many reasons why the spring could be loose, for example: it could be due to a bad spring lot, the wrong spring installed, poor installation of the spring, the clip that holds the spring could be bad, and so on.

So we ask the question the third time, "Why is the spring loose?" This time we find out that the wrong spring was installed. At this point we may be at the root cause of the problem except that we do not know why the wrong spring was installed. Some of the reasons why the wrong spring was installed could be: the bill of materials was wrong, the wrong spring was pulled from the stockroom, the supplier mislabeled the springs, or the wrong springs were ordered. Because there are still unanswered questions, we have not yet reached the root cause of the problem.

So we asked the question a fourth time "Why was the wrong spring installed?" This time we find out that purchasing ordered the wrong spring. As you continue to peel away each layer of the onion, you get closer to the root cause of the problem. However, until you understand why purchasing ordered the wrong spring you will not have reached the root cause the problem.

So we asked the question one final time "Why did purchasing order the wrong spring?" This time you find out that the spring needed for the assembly was unavailable and that a distributor recommended that an equivalent spring be purchased. The spring turned out not to be an equivalent. Now that you've reached the root cause of the problem you can begin to explore solutions that will permanently correct the problem. For example, engineering must approve all part substitutions and manufacturing must test all alternative parts to verify compliance prior to using the alternative component. Since the purchasing agent was told that this spring was an equivalent part, it is possible that engineering could make the same mistake. To prevent this from happening in the future, all substitute parts must be verified in the system before they can be used in the production line. You can also use HALT to evaluate substitute part acceptability. (In tight situations, you may still have to order the alternative component in order to meet production schedules, but the evaluation of the alternative component must be done prior to placing it on the production line.)

Finally, identify past customer complaints, safety warnings, and product recalls along with the root causes associated with each problem. Pareto the list on the basis of severity and occurrence.

Combine all the lessons learned from past problems into a Pareto of severity. This list will be used in the design concept phase to describe how past problems will not be repeated in the new product.

13.5 RISK MITIGATION

In Chapter 5, we presented a process to manage the risk issues inherent in product development. Technology, component, process, supplier, and safety are all areas in which there can be significant reliability risks. If these issues are left unresolved, they will probably surface later as reliability problems found in manufacturing, test, or by the customer. The best way to eliminate this problem is through risk mitigation. A risk mitigation program not only removes reliability issues in design, it also prevents delays in product development associated with risk issues that require redesign late in the development cycle.

In Chapter 5, we showed how to capture and document the risk issues. All of the functional groups are responsible for developing risk mitigation plans. The initial risk mitigation plan is developed at the product concept phase. The mitigation plan is updated in each phase of the product development cycle to reflect the risk mitigation progress and add new risk issues that were discovered. Before the end of the product concept phase, a meeting is scheduled for all the functional groups to present their risk issues. The meeting is a formalized event (i.e., "Risk Mitigation Meeting") that is part of the product development process.

The risk mitigation meeting is designed to review the significant risk issues and agree on the risk issues, costs, and resources to mitigate these issues; and that there is sufficient time to mitigate before shipping customer units. The development cost at the product concept phase is small and increases significantly in the design phase. The risk mitigation meeting is a useful way for management to determine if a program should proceed with funding to the next phase or be cancelled because risk issues are unlikely to be resolved in time to meet the product market window. How many products, from your own experience, were canceled before being completed? Were they cancelled because the product was too far behind schedule, too costly, or technical issues were not resolved? Where in the product development cycle were these projects cancelled? Usually, it is late in the development program after the development budget has been exceeded and a litany of design issues still need to be resolved. The programs from which there is little hope of resolving critical risk issues are identified early in the program and terminated well before excessive amounts of capital and personnel resources are wasted. If the program is vital to the success of the business, then resources can be allotted to determine the cause of the problem and recommendations made to remedy the situation.

Plan the risk mitigation meeting towards the end of the product concept phase. If there are only two groups involved in product concept, marketing, and engineering, then the meeting will be short. However, as we have shown in Table 13.1, it is a good idea to include as many of the functional groups as possible. In particular, reliability, customer service/customer support, manufacturing, and test should be included in the concept phase risk mitigation meeting. At least a week before the risk mitigation meeting, a risk mitigation package should be put together containing the risk issues identified by the participating functions. The meeting will run smoother if the information is circulated with sufficient time for each organization to review the issues, understand the problems, and discuss them before the mitigation meeting. The package should include all the material that will be presented in the meeting. A sample product concept phase risk mitigation form is shown in Table 13.2. By starting

Table 13.2   Product Concept Phase Risk Mitigation Form

Date: 07/01/2001

Product name:

Owner: Mr. Jones

Reliability ICM – product concept phase

Item no.	Investigate		Communicate		Mitigate				
	Identify & analyze risk (1)	Risk severity (2)	Date risk identified (3)	Risk accepted Y/N (4)	High level mitigation plan (5)	Resources required (6)	Comp date (7)	Success metric (8)	Investigate alternative solutions? (9)
1									
2									
3									
4									
5									
6									
7									
8									

Risk mitigation signoff: _____

risk mitigation in the product concept phase, you significantly improve the likelihood that the product will be developed on time and within budget.

13.5.1 Filling Out the Risk Mitigation Form

The risk mitigation form has nine parts. We have tried simpler versions of the form but always end up going back to this greater level of detail. Details on filling out the form follow:

1. *Identify and analyze risk*
There are two parts to investigating risk. The two parts are identifying the risk and analyzing the risk. The first part, identifying the risk issues, is usually the hardest. Determining the risk issues this early in the program can be a real challenge. The product concept phase risk issues will be technology and reliability related. Since the design has not started, the risk issues may be few and not detailed. The following questions can help you identify product concept risk issues:

Places to look for Technology risk issues:

- Are there new technologies required?
- Are there new technologies at the bleeding edge, state of the art, or leading edge of technology?
- Are new processes (manufacturing, rework, and test) required for this technology?
- Is there little or no information published regarding these technologies?
- Is there only one supplier for this new technology?
- Is the new technology not available yet commercially?
- Any issues critical to program success?

Places to look for Reliability risk issues:

- FRACAS report and Pareto of lessons learned from past programs,
- the results from the external VOC,
- file of past product recalls and safety warnings.

Once the risk issues are identified, you can begin to analyze them. Identifying the risk answers the question, "What are the critical risk issues and why are they critical?" In analysis, we answer the question "What needs to be done to eliminate the risk?" Things to consider when analyzing the risks are as follows:

- Are there special skills, equipment, or resources required that either do not exist within the company or are unavailable?

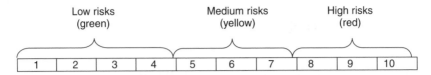

Figure 13.1 Risk severity scale

- What testing is required?
- What is the impact to the program?

2. Risk severity

It is vital to associate with each risk a level of severity. This can help in managing the vital resources to mitigate the most severe and critical risk issues first. To differentiate severity, use a numbered scale, for example, one to ten. Ten represents the most severe risk and one is the least severe risk, see Figure 13.1. The scale is further differentiated into bands of high, medium, and low risk. The scale can be color coded to give greater visibility to higher risk. High risks are red, medium risks are yellow, and low risks are green.

3. Date risk is identified

Documenting when risk issues are identified is useful while reflecting back to determine how to improve the process. It is important to know if critical risk issues are being identified late in the product development cycle. Keeping track of these dates is also useful in monitoring your success to identify risks earlier in subsequent programs.

4. Risk accepted

Each risk issue is reviewed at the risk mitigation meeting. A decision is made then to accept the risk or reject the risk issue. Use this block to keep track of the risk issues that will be mitigated. A follow-up meeting may be needed if there are risk issues that are not accepted.

5. High-level mitigation plan

Detail the activities that will take place to mitigate the risk. Include any outside contract or service provider work that will take place. The high-level plan should have sufficient detail to show progress.

6. Resources required

Identify at a high level of detail the resources required to mitigate the risk. The resources should include manpower (i.e., 10 months, 2 people) and capital resources (include equipment, testing and evaluation services, consultants, and other associated R&D costs). Once the risk issues are accepted, the high-level resources are to be inserted into the work breakdown schedule and the departmental capital budget forecast.

7. Completion date

Enter the expected completion date to mitigate the risk. The completion date must support the first customer ship date. Note if this date has slipped, especially if it has slipped several times. This is a strong indication that the risk is high and/or the resources to mitigate the risks are low.

8. *Success metric*

The success metric is one of the most important and often overlooked factors in mitigating risk. All too often risk mitigation activities are launched without clear direction of what results are needed for success. This can often lead to either doing much more than what is needed or doing the wrong activities to mitigate risk. By having clearly defined success metrics, the team that is launched to mitigate the risk will be focused on those activities that support the success metric. By getting the other functional groups to agree on the success metrics up front, there will be no disconnects discovered later in the product development.

9. *Investigate alternative solutions*

For any risk on the critical path that is vital to program success, or if the risk is between a level eight and ten (red high level), there should be a contingency plan in place to mitigate the risk. Often the contingency plan can have a significant impact on the product design or product delivery date. Because of this, it is important to include a date when a decision needs to be made to launch the contingency plan activities.

13.5.2 Risk mitigation meeting

The last activity that takes place before proceeding to the design concept phase is the risk mitigation meeting where all the risk issues are reviewed. In the risk mitigation meeting, each group presents their risk issues with the objective of achieving the following results:

1. communication of all significant technology risk issues,
2. agreement on the severity of the risk issues,
3. agreement on the strategy to mitigate the risk issues,
4. agreement on the metric for successful mitigation of risk,
5. agreement on the resources required and availability of the resources to mitigate risk,
6. agreement on the timeline required to mitigate risk,
7. agreement on the need for the risk and/or the need to pursue an alternative solution.

Figure 13.2 ICM sign-off required before proceeding to design concept

The risk mitigation meeting requires sign-off on every risk issue. By requiring sign-off on every issue, the ICM process acts like a project gate as to where the program will proceed, only if the risk issues are manageable (Figure 13.2). Project funding in the form of capital and staffing for the development of the next phase is contingent on obtaining sign-off by the senior management on the risk issues.

14

Design Concept Phase

Previously, in the product concept phase, the product requirements are defined on the basis of market-driven product features – cost, forecasted demand, target customers, and business fit. This is where we produce a set of design requirements, and possibly, high-level system architecture. Once these product requirements have been defined, a design concept must be developed to meet these needs. The design concept phase uses the product requirements to develop lower-level design architecture. Upon completion of the design concept phase, the specifications for the outline dimensions, weight, input and output (I/O), power, cooling, and so on is determined.

Decisions made in the design concept designate what type of components, materials, and technologies are required to design the product. These decisions have a significant impact on the product cost, development time, design complexity, manufacturability, testability, serviceability, and reliability of the product. At the completion of the design concept phase, about half of the product cost is defined (see Figure 14.1). Product cost is a significant factor in profitability. This is where the design concept phase is used to ensure that the product cost goals are obtainable.

There are five reliability activities that take place in the design concept phase (Table 14.1). These are as follows:

1. Set reliability budgets for subsystems and board assemblies

2. Define reliability design rules

3. Revised risk mitigation

4. Set capital and personnel reliability budgets (reliability activities are included in program development schedule)

5. Decide risk mitigation sign-off day.

Improving Product Reliability: Strategies and Implementation. Mark A. Levin and Ted T. Kalal
© 2003 John Wiley & Sons, Ltd ISBN: 0-470-85449-9

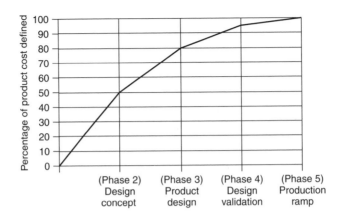

Figure 14.1 Opportunity to affect product cost

Table 14.1 Design Concept Phase Reliability Activities

	Design concept phase	
Participants	Reliability activity	Deliverable
• Reliability engineering	1. Define lower-level reliability design goals.	1. Subsystem and board-level reliability budgets (MTBF), service and repair requirements (MTTR and availability), and product useful life and use environments.
• Marketing		
• Design engineering (electrical, mechanical, software, thermal, etc.)	2. Define reliability design rules.	2. Identify guideline requirements for DFR, DFM, DFT, DFS, etc.
• Manufacture engineering	3. Risk mitigation revised. Include internal reliability VOC (manufacturing and test) and new technology issues.	3. Pareto top reliability issues with recommendations. Risk mitigation plan updated with changes.
• Test engineering	4. Reliability budget (capital & personnel) and reliability activities are included in program development scheduled.	4. Reliability expenses are budgeted and reliability activities (risk mitigation, FMEA, HALT, HASS) scheduled into project timeline.
• Field service/customer support	5. Review status and agree on each risk issue and mitigation plans. Risk mitigation meeting.	5. Risk mitigation meeting and agreement to proceed to next phase.
• Purchasing/supply Management		
• Safety & regulatory personnel		

Note: DFM: Design For Manufacturing; DFR: Design For Reliability; DFS: Design For Service (and maintainability); DFT: Design For Test; FMEA: Failure Modes and Effects Analysis; HALT: Highly Accelerated Life Test; HASS: Highly Accelerated Stress Screens; MTBF: Mean Time Between Failures; MTTR: Mean Time To Repair; VOC: Voice Of Customer.

The product concept team is still typically small, consisting of key design personnel from marketing, engineering, and reliability. The development cost and staffing resources expended to this point are also relatively small. Once the product is out of the concept phase and in the design phase, the staffing and capital resources required increase considerably. Because there is a serious commitment of staff and capital resources required in the design phase, there must be a high level of confidence that the development program will be successful. The reliability activities in the design concept phase primarily focus on capturing the risk issues that can sidetrack a development program. By capturing these issues and developing mitigation plans, an assessment can be made at the end of the product concept phase to either proceed with the product design or stall its progress until key risk issues can be resolved.

14.1 SETTING RELIABILITY REQUIREMENTS AND BUDGETS

After the product has transitioned from a product concept to a design concept, guidance will have to be provided by the reliability team detailing the reliability design requirements. These requirements can include the following:

1. Product use environment
2. Product useful life
3. Subsystem and PCBA reliability budgets
4. Service and repair.

Requirements for product use environment   The customer use environment is defined either for the typical customer or for the extremes of customer use. Setting the requirements for the typical customer will optimize the design for product cost. This can, however, result in higher customer failure rates for those customers operating at the environmental extreme. Designing for the extremes of customer use will result in a more complex and costlier product, however, it will be more reliable (this is evident in hi-rel military products). The product may have a larger potential market but will sell for a premium. This higher product price will impact sales. It is best to design the product for this optimum customer base, often referred to as the market "sweet spot." Once the environment is decided, these requirements become the product's environmental specifications.

Environmental requirements include both operational and nonoperational specification. The nonoperational requirements consider storage and transportation. Some shipping and transportation environmental concerns are as follows:

- If the product is shipped overseas will it be exposed to salt air and for how long? For what length of time will the product remain on a loading dock? Will it be exposed to rain, snow, sleet, or dust? Air transportation can avoid many of these problems.

- What type of vibration shock levels and vibration frequency spectrums will the product be exposed to? Studying the shipping environment to design a proper shipping container/method can avoid the "dead on arrivals" or so-called *out of box failure* complaints by customers.

- What environments will the product be stored in? Will it be shipped to a desert area where cargo temperatures can be excessive? Will it be stored in a humid climate? How long can it be stored in any of these environments?

The operational limits are defined for the environments where the product is expected to normally operate. Operational limits can include operating temperature range, vibration range, and frequency spectrum, duty cycles, line voltage/frequency variation, drop or shock test, moisture exposure or water immersion test, corrosive environments, and so on. Once defined, test plans can be developed to validate the design.

Product useful life requirements   It is important to define what the useful life requirements are for the product. The useful life is the period of time when the failure rate turns from a random event to a predictable event based on normal wear (see Figure 14.2). Most product developments proceed without any direction regarding the required useful life of the product. If it is not defined up front, there will be different perceptions about the useful life requirements amongst the design team. A common example of wearout would be a light bulb with a 2,000-h life. The 2,000 h defines the useful life of the light bulb. It can be easily repeated and if a test was performed on a large enough sample, similar results would be achieved each time.

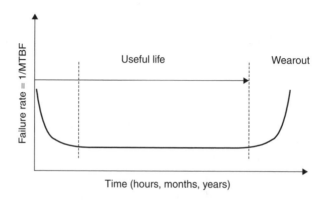

Figure 14.2   The bathtub curve

Useful life design requirements are important because they impact design cost, architecture, and complexity. Setting the requirements too stringently will make the product more costly and prolong design development. These requirements are also needed to properly select components in design. The requirements are market-driven, supported by market research, and specific to the product being developed. The useful life requirements can be designed to meet the sweet spot in the customer market, or for the extremes of expected customer use.

The useful life of a product is typically defined by some amount of customer use time. The product development team may define the system operating useful life in years, that is, seven years.

The useful life can be defined in different ways for the same product. Some ways to define useful life requirements are

- minimum number of failure-free mating cycles (connectors, removable accessories, etc.),
- minimum number of on/off cycles (relays, hard drives, lubricated bearings, etc.),
- minimum number of running hours (fans and motors, pumps, seals, filters, etc.).

These other requirements correlate back to the system requirements. There is consistency between the different useful life requirements even though they are defined differently. Consider a product with a seven-year useful life. The product has an accessory that is removable and marketing research expects the accessory to be removed five times a week. The useful life for this accessory can have a minimum number of failure-free mating cycles defined for it. In this case it would be:

$$\text{Cycles(min)} = (7 \text{ years}) \times (52 \text{ weeks/year}) \times (5 \text{ cycles/week}) = 1,820 \text{ cycles}$$

Once these requirements are defined, the next task is to verify down to the component level that every part complies with the useful life requirement. This may appear to be an overwhelming task, but it is not. Many of the components will be grouped and evaluated by their device and component package type. For example, all carbon composite surface mount resistors can be grouped by package type (i.e., 0805, 0603, 0402, etc.). Some component manufacturers will specify the useful life. The useful life of a part is often a function of its environment and use conditions. An electrolytic capacitor manufacturer, for example, will specify the maximum useful life of their capacitor at a particular temperature, that is, 2,000 h at 105 °C. The temperature that the product operates may be only 45 °C. This will increase the expected useful life of the capacitor. For many electrolytic capacitors, there is a rule of thumb which states that the useful life increases twofold for every 10-degree temperature decrease. Therefore, we should expect the capacitor's useful life to extend to 64,000 h.

Those components that will wear out before the useful life of the product occurs need to be designed for maintenance and service. By setting requirements upfront, there is a process to capture and address those components that do not meet the customer-expected life requirements. Designing the parts that will wear out before the end of the product life to be *easily replaceable* will help reduce service cost and improve customer satisfaction.

Subsystem and Printed Circuit Board Assembly (PCBA) reliability budgets In the previous phase, system reliability requirements were defined. For example, the system needs to have a 10,000 MTBF. In the design concept phase, the system reliability budget is tiered down to the subsystem and circuit board level. Reliability budgets are not defined to the component level for several reasons. First, in the design concept phase, the components required to design the product are only partially known. Second, and more importantly, one of the purposes of the reliability budget is to provide guidance to the design team as they begin to design the subsystems, circuit boards, and interfaces. The reliability budgets are used to identify reliability risk issues. Because resources and time are limited, the reliability budgets can steer the reliability activity in a direction where it will most benefit the product. The reliability budgets can also provide guidance regarding which component is more appropriate to achieve the reliability goal and for determining the need for redundancy to improve reliability.

The system reliability requirements were defined in the product concept phase, an example of a system reliability budget is shown in Figure 14.3. The system reliability budget can also include requirements for mean repair time, serviceability, availability, and product life requirements. Once the architecture is defined, assign reliability budgets for all the subsystems, circuit boards, and mechanical assemblies that make up the system or product. The sum of all the subsystem reliability budgets must be equal to the system reliability defined earlier.

If the reliability is defined in MTBF (Mean Time Between Failures, in hours), then you must add up the reciprocal of each MTBF. The MTBF budget numbers for each of the subsystems cannot be added directly. If the reliability for the subsystems is defined in FITs (Failures In Time, per billion hours), then they can be directly added to determine the equivalent system FIT rate. FIT numbers can be converted to MTBF by taking the reciprocal of the FIT number and

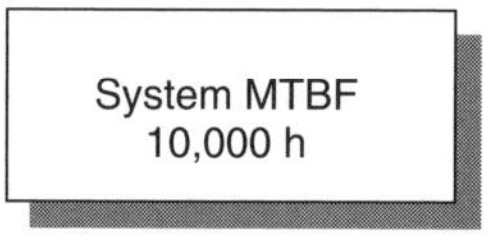

System MTBF
10,000 h

Figure 14.3   System MTBF requirement

How is Calculated

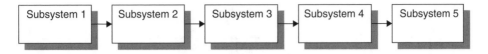

The above system is comprised of five subsystems

Figure 14.4   Subsystem MTBF requirement

multiplying it by one billion to arrive at MTBF numbers in hours.

$$\text{MTBF} = \left(\frac{1}{\text{FIT}}\right) \times (1 \times 10^9)$$

An example of how the subsystem MTBF budgets can be added together to equal the system reliability budget in shown Figure 14.4. Suppose that you are working on a new product development whose system reliability budget (stated in MTBF) was defined in the product concept phase to be 10,000 h. In the design concept phase, the system architecture shows the system to consist of five major subsystems. The reliability organization is then tasked to define the MTBF budgets for the five major subsystems. The MTBF budgets represent reliability's best guess at what the MTBF requirements need to be for the different subsystems. If the widget is a completely new product, then the reliability budgets may be nothing more than an educated guess for each of the different subsystems. However, if the new product has many similarities to previous products there is a database of knowledge regarding the previous product's reliability, broken down into their subsystems to the board level. A Failure Reporting, Analysis, and Corrective Action System (FRACAS) system will contain information regarding the previous product's reliability, which can be broken down into its system and subsystem components. To learn more about MTBF refer to Chapter 6 and Appendix B.

System MTBF

$$= \left(\frac{1}{\left(\frac{1}{\text{MTBF}_1}\right) + \left(\frac{1}{\text{MTBF}_2}\right) + \left(\frac{1}{\text{MTBF}_3}\right) + \left(\frac{1}{\text{MTBF}_4}\right) + \left(\frac{1}{\text{MTBF}_5}\right)}\right)$$

$$= \left(\frac{1}{\left(\frac{1}{17,000}\right)+\left(\frac{1}{28,000}\right)+\left(\frac{1}{350,000}\right)+\left(\frac{1}{940,000}\right)+\left(\frac{1}{650,000}\right)} \right)$$

$$= 10,000$$

How does one determine the budget numbers? The best way is to review the past performance of similar subassemblies, the key contributors to this performance, and the expected result if these issues were removed in the new product.

The first time you set the reliability budget (for each subsystem) will always be the hardest. With each new program it will get easier. This is because you can leverage the lessons learned from the previous programs and have a better idea as to what the reliability budgets should be for the subsystem and circuit board levels. Appendix B discusses MTBF calculations in more rigorous detail.

Service and repair requirements Service and repair requirements can be as simple as the maximum time allowed to replace a faulty device. It can also include statements regarding the modularity of devices that are expected to be replaced during normal use. Can special tooling or equipment be used during replacement? It is a good idea to design out the need for special tooling and equipment that is required for normal service. Finally, will there be any requirements to prevent the incorrect installation or alignment of replacement parts?

14.2 DEFINE RELIABILITY DESIGN GUIDELINES

The reliability staff is tasked to ensure that the new design will be reliable. However, reliability engineers do not design products. Designers do. Therefore, it is important that the design team have a set of design rules that they can follow for making the right decisions for a reliable product. This can be achieved through a set of design rules for product reliability. These rules are referred to as the Design For Reliability (DFR) guidelines for product design. Adhering to the DFR guidelines should be one of the requirements agreed to in the design concept phase. The DFR guidelines help the designer to make the right decisions. These guidelines are broad-based and cover all aspects of the product design. They must incorporate all of the reliability lessons learned from previous programs. The DFR guidelines should be part of a design checklist or other similar means used to ensure that the design is ready for manufacturing. The reliability design rules change over time because of new problems, new technologies, and new capabilities. The best way to communicate these changes to the design team is by periodically providing DFR classes to the organization. Provide training sessions as an opportunity to discuss new reliability guidelines, changes to previous guidelines, and to solicit feedback about how these guidelines will impact design engineering. Use the training as an opportunity to find out if there are gaps in the DFR guidelines or if reliability design issues are not being applied.

The DFR guidelines should be easily accessible to the design team. One way to achieve this is by placing the guidelines on the company intranet where access is readily available. Placing the guidelines on the intranet empowers the design team to make the right decisions without reliability involvement on every problem. Each reliability guideline should state the problem clearly and specifically. The DFR guidelines should not be so general that they cannot be applied in design. The guidelines should state the conditions in which reliability is a concern. Each DFR guideline should define the impact it has on product reliability. Finally, each guideline should call out ways to improve the reliability via methods like derating or reducing the operating temperature, and so on. Each DFR guideline should include suggested alternatives. (It is good to provide what not to do; it is better to also provide alternative solutions.)

14.3   RISK MITIGATION IN THE DESIGN CONCEPT PHASE

The Identify, Communicate, and Mitigate (ICM) process is repeated in each phase of the product development to identify new risk issues and to report status of the risk mitigation progress from the previous phase. There was very little known in the product concept phase regarding the design of the product. Because of that, the risk issues identified were mostly customer-focused. In the design concept phase, details begin to emerge regarding what the structure of this design will look like. As more details begin to develop about the design of the product, new risk issues will emerge. Therefore, the risk mitigation process is repeated in the design concept phase.

These new risk issues, if unresolved, will cause significant delays in product introduction. The delay results because risk issues, if unaddressed, will surface later in the program, which will require resolution. The later in the program the risk issues are addressed, the more significant the impact they will have on the program. Finding major problems in prototype is costly and delays product launch due to the time required redesigning and implementing fixes. Ideally, risk issues are identified and mitigated before design validation.

14.3.1   Identifying Risk Issues

Risk mitigation requires a 180-degree approach. Risk issues can be found by reflecting back on past problems and looking ahead to see what new issues pose

Figure 14.5   180° of risk mitigation

significant risk (Figure 14.5). In the product concept phase, we reflected back on the external lessons learned. In that phase we captured the lessons learned from customer feedback, FRACAS, complaint files, product recalls, and safety warnings. Using a Pareto chart, the top issues were identified and transferred to a risk mitigation plan. These issues were then inserted into a high-level risk mitigation plan and were tracked to closure. These risk issues are the lessons learned from the past. The process of capturing these issues, developing plans to alleviate them, and tracking the progress made on these plans is risk mitigation. Incorporating the external lessons learned into the concept requirements will help ensure that they are not repeated.

On reflecting back, the external lessons learned were captured into the product concept. In this phase we capture the internal lessons learned from manufacturing, test, design, and reliability. The internal lessons learned are added to the risk mitigation plan as new risk issues.

14.3.2   Reflecting Back (Capturing Internal Lessons Learned)

The most effective way to reflect back is through a review of the (hopefully) documented lessons learned from the past. If there is no mechanism to capture the lessons from the past, you can expect to continually repeat the same. This is especially true during a growth phase when a product is ramping up and staff is being added to meet demand. A comprehensive review of the lessons learned from the previous program includes an examination of what went wrong, what didn't work, and what worked well. This is the same activity performed in the product concept phase except that it is focused on identifying internal lessons learned. There are four areas to focus on when capturing the internal lessons learned. The four lessons learned come from:

1. manufacturing and rework,
2. test,
3. component and supplier issues,
4. reliability.

The best way to capture the lessons learned from previous programs is by documenting the issues from the past program one year after it has reached volume production. Each of the functional groups should be responsible for documenting the issues that were significant during product development, early production, and product ramp. Capturing the issues that impacted a program while in production minimizes the effect of people forgetting things over time or those who have left the company. Be specific about the impact of the problems in the documentation. The problem may have caused a low first-pass yield in test; including the yield expected and the actual yield helps to quantify the magnitude of the problem. Some of the common ways to define the yield are the Parts Per Million (PPM) and process control charts. Establishing required PPM rates for test and process control limits for production will aid in identifying problems early in production. If your company has an Intranet, then the lessons learned from the previous programs can be posted and made available for everyone.

14.3.3   Looking Forward (Capturing New Risk Issues)

There is also risk in the unknown. Looking ahead is the process used to identify new risk issues and is often based on past experience. For example, if Application Specific Integrated Circuits (ASICs) have a history of design-related failures, then ASICs should be a risk mitigation issue. Looking forward requires identifying technology, supplier, manufacturing, and test and design issues, which pose unique and challenging problems difficult to resolve. These risk issues, if not mitigated, will probably manifest themselves later as reliability problems. The following questions will help in identifying new risk issues:

Places to look for design risk

- Are new technologies required?
- Are new component packages used?
- Are any aspects of the design approaching or exceeding technology limits?
- Are complex ASICs or hybrids required?
- What parts are custom or nonstandard?
- Will you be approaching or exceeding the stress limits of any component, packages, or designs?
- Are there tight electrical specifications?
- Are there tight mechanical tolerances and tolerance stack-ups?
- Is excessive weight an issue?
- Are there material mismatch or incompatibility issues?
- Are there corrosion or other chemical reaction issues?
- Are there unique thermal issues?
- Are there any premature component wearout concerns?

- Is there a history of problems with similar parts or device types?
- Are there components whose Electrostatic Discharge (ESD) sensitivity exceeds capability? There are three ESD models to consider (human body model, machine model, and charge device model) on the basis of use conditions.
- Does the component count exceed capabilities?

Places to look for manufacturing risk

- Are new processes required?
- Are new packages required?
- Are there high-pin density parts that exceed present process capabilities?
- Does the new design push the present limits of manufacturing capability?
- Does the component count exceed manufacturing capabilities?
- Does the design require high part density (parts/square inch)?
- Will there be any special handling requirements or concerns?
- Are there new components requiring small package sizes that exceed present capabilities?
- Are there new components with large packages that exceed present capabilities?
- Are there new components with heavy packages that exceed present capabilities?
- Are there new components with high defects per million opportunities (DPMO) packages that exceed present capabilities?

Places to look for supplier risk

- Are there new custom parts that pose special risks?
- Do the new designs require Printed Circuit Board (PCB) technology that exceeds present supplier limits (e.g., size, weight, lines and gaps, number of trace layers, core size, copper weight, material selection, and material compatibility)?
- Is there a history of problems with any of the proposed suppliers?
- Will new designs require supplier qualification?
- Does any of the supplier's small size pose special risk issues (supply capacity, financial stability, etc.)?
- Are there any critical components that cannot be dual-sourced?

Places to look for test risk

- Are there new components or new packages that cannot be tested due to access restrictions?

- Do high-pin count/density component packages pose special test challenges?

- Does the product push the limits of testing capability?

- Is the PCB part count high enough to exceed test node capability?

- Are new component packages too small to probe?

- Are there new component packages that cannot be tested because of access?

- Are there hybrids or other parts that cannot be fully tested?

- Are there parts with high PPM rates that will result in low first-pass yield?

Use the risk mitigation tool to provide the most effective way to capture risk issues and assign values identifying the magnitude of the risk. Each of the risk issues is assigned a number that represents the degree of risk severity. The risk issues with the highest severity ranking need to be resolved first. All high-risk issues should have plans for an alternative solution if they cannot be resolved before first prototype. These risk issues will have a significant impact on the ability to meet the time-to-market goals. These risk issues, if unresolved, will probably return as reliability problems once the product goes into full production.

Once the risk issues are agreed to by the team as being necessary, risk mitigation plans are put in place with timelines defining when the risk will be mitigated. Finally, the risk mitigation plans identify the metrics that will be used to determine successful mitigation of the risk issue. Defining the metrics for successful risk mitigation is often overlooked. Too many programs expend significant amounts of time and resources toward resolving problems. The work performed to resolve the risk issues is often very good but not focused around what is needed to mitigate the risk. In other words, a lot of testing and analysis is performed, but because it was not focused around the problem statement and success metric, the problem is not rectified. By clearly defining what is required to define successful risk mitigation, the problem statement is more easily resolved.

One way to ensure that all the risk issues are identified early in the design phase is to have a cross-functional risk mitigation meeting to review the design concept and identify risk concerns. The risk mitigation process begins early in the design concept phase with a kickoff meeting to discuss the design concept and to solicit feedback from the cross-functional teams. It is then the responsibility of each of the functional groups to begin documenting and tracking risk issues. Periodically through the design concept phase, the cross-functional team should meet to discuss the status of the risk issues, identify new issues, and discuss alternative solutions.

14.4 RELIABILITY CAPITAL BUDGET AND ACTIVITY SCHEDULING

An activity often overlooked in reliability is the upfront planning that needs to take place. Planning is crucial. By having the reliability activities scheduled into the project timeline, common arguments like *"These reliability activities will prevent us from meeting the market window," "I cannot free-up the resources you need because everyone is working on items on the critical path," "It is too late in the program to fix anything we find. It would have been nice to have done this earlier in the project"* are avoided. There is no excuse for not planning these activities into the project timeline. Scheduling ensures that these activities take place in the appropriate stage of the product development cycle, and not when the product is in production. You can be assured, any finding from HALT and FMEA after a product is in production, will in all probability, not be implemented.

Scheduling the reliability activities into the project timeline also confirms the commitment of the management to these activities, and the importance of these activities to the program. Participation is not optional. A typical FMEA will take between a half to a full week to perform. This does not include the time required to fix problems found. Scheduling a week for this activity will usually suffice. Complex systems should be broken into smaller parts so that they can be performed in a week. Remember that this is a concurrent activity that will require other functional groups to schedule this activity into their project timeline.

FMEAs should also be planned and scheduled even if an outside supplier will perform them. Subsystems and custom designed parts that are purchased by an outside supplier also require an FMEA. The FMEA should be a joint effort between the supplier and the customer. If the outside supplier is unfamiliar with an FMEA, then additional time must be allotted for training. The training can be performed all at once in the beginning or before each of the three major parts (functional block diagram, fault tree, and failure modes and effects analysis form). If the supplier is a significant distance away, it may not be practical for the participants to all be in the same room for the FMEA. We have had significant success using Microsoft NetMeeting® to tie in different facilities and keep everyone focused. There can be issues with computer firewalls that may need to be worked out, but the process can be performed very effectively with remote sites.

The other significant activity that has to be scheduled is HALT. Allow a week to two weeks for the HALT activity for each subassembly. A second HALT test is usually performed to verify the design fixes and to ensure that no new failure modes were designed into the product as a result of the design improvements. The HALT scheduling is critical because there are many activities that take place before the HALT test is performed. HALT requires a significant amount

This whole section is good

of up-front preparation. Before the HALT test can be performed the following things are needed:

- HALT test plan
- Stress level limitations
- Mechanical fixturing
- Input/output cabling
- Instrumentation
- Test software ready
- Product test hardware built and debugged
- HALT Team identified (the team must be there during the HALT testing as the test evolves for optimum efficiency).

These activities often take place months before the actual HALT test is performed. Scheduling these activities far enough in advance will ensure that when it is time to test hardware, you will be ready. Last minute planning guarantees sputtering starts and lost time due to oversights, and so on. The HALT planning activities are covered in detail in Chapter 15.

Finally, there will be capital expenses associated with the reliability activities that should be budgeted into the program development cost. The majority of the expense is for HALT testing. The HALT test typically requires four working test devices, three of which are destructively tested. The fourth is a golden unit for trouble shooting which can be returned for other uses or sold. However, the material cost for the three boards can be significant depending on the product. In addition, there can be other costs associated with HALT, that is, fixturing, cabling, test hardware/equipment, electric power, and liquid nitrogen used to run the chamber. If the HALT testing is performed by an outside test service, this too will need to be budgeted. HALT testing may identify a bad part that requires further failure analysis. Thus, it is wise to plan and budget for some amount of outside failure analysis work and possibly accelerated life testing.

By doing a thorough job in planning for the reliability activities, the risk of roadblocks surfacing is minimized. The planning phase will lay out all the activities that need to take place, their estimated duration and time frame when these events need to occur. There will be no surprise of its impact to the program and the ability to meet the marketing window. The resources needed to perform these activities will be budgeted into the program early on so that the funds are available when needed. If there are capital constraints, then adjustments should be made early on so that they do not affect critical assemblies.

Figure 14.6 The ICM is an effective gate to determine if the project should proceed

14.5 RISK MITIGATION MEETING

At the completion of the design concept phase, the entire team understands the program risk issues and should be working cross functionally toward risk mitigation. The majority of the program risk issues should be identified by the completion of the concept phase. A new synergy will form and the team will work more efficiently to eliminate the shared program risk. The team will better understand and appreciate how the decisions made in the design concept phase will affect product reliability, cost, and time to market.

Remember, toward the end of the design concept there is a planned risk mitigation meeting to review all significant risk issues. The risk mitigation meeting should occur before the design is allowed to proceed to the design concept phase. The risk mitigation meeting acts as a project gate where a decision is made to proceed to the next phase of product development (Figure 14.6). If the risks are not being adequately managed, then the project success can be at risk. By the end of the product concept phase, only a small percentage of the total program development cost has been spent. Proceeding past this point requires significant capital investment and resources. Risk management is an effective tool to determine if the program is ready to proceed further.

The structure for the risk mitigation meeting is the same as that used in the product concept phase. Attendance is required at this meeting. Prior to the meeting, a risk mitigation issues package is put together for all participants. The package includes all the material that will be presented in the meeting. The risk mitigation meeting reviews all significant risk issues to determine if they will be mitigated in advance of first customer ship. The risk mitigation meeting is scheduled toward the end of the design concept phase.

14.6 REFLECTION

The last activity before the concept phase ends and the design phase begins is reflection. In the reflection step you look back at the concept phase to see what worked and what didn't. Capture the lessons learned early so that they are not forgotten and destined to be repeated. Document the findings and recommendations so that they may be incorporated into future programs. Reflection allows for continuous improvement of the reliability process.

15

Product Design Phase

15.1 PRODUCT DESIGN PHASE

Now that the concept, requirements, and architecture for the product have been completely defined, the design of the product can commence. In the product design phase, everything required to produce a working prototype is developed. At the end of this phase there will be a complete product documentation package. This will include the schematic, theory of operation, outline drawing, Bill Of Materials (BOM), software, assembly, and mechanical drawings. There will also be a working prototype suitable for design validation, which will be performed in the next phase of product development. The decisions made by the end of this phase will determine the product cost, design, manufacturing, test, and service complexity and will also determine how difficult it will be to ramp production. Unless the product is completely redesigned, 80 to 90% of the product cost is determined. Cost down efforts to reduce product cost here are usually limited to reducing material cost because redesign at this point is not cost effective.

Product cost is emphasized because there is a cost associated with reliability. The cost equation works two ways. First, there is an added material cost associated with using higher reliability components and adding redundancy. The added cost should be evaluated against the improvement expected in reliability and what the market is willing to pay for increased reliability. There is also a cost saving associated with designing the product right the first time. Typically, early in the product launch, there is an extensive number of engineering design changes to the product (refer to Chapter 1). Design changes at the end of the product development increase the cost to produce the product and can reduce product reliability when there is extensive rework and retrofit required.

The design phase has two parts (Figure 15.1), a product design phase and a design validation phase. The majority of the engineering design activity takes

Improving Product Reliability: Strategies and Implementation. Mark A. Levin and Ted T. Kalal
© 2003 John Wiley & Sons, Ltd ISBN: 0-470-85449-9

Product design phase

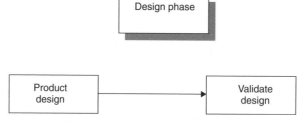

Figure 15.1 The first phase of the product life cycle

Table 15.1 Reliability Activities for the Product Design Phase

Participants	Product design phase	
	Reliability activities	Deliverables
• Reliability	1. Reliability estimates developed for all lower level assemblies. Identify all items, early wearout items.	1. Reliability estimate spreadsheet and a Pareto by component and subassembly highest failure rate items. Pareto early wearout items for service and maintenance strategies.
• Marketing		
• Design engineering (electrical, mechanical, software, thermal, etc.)	2. Implement risk mitigation plans.	2. Risk mitigation meeting and agreement to proceed to next phase.
• Manufacturing engineer	3. Apply reliability design guidelines (DFM, DFR, DFT, DFS).	3. Checklist or review that design guidelines has been followed and variances are acceptable.
• Test engineer	4. Perform design FMEA.	4. Completed FMEA Spreadsheet (Table 7.1) and closure on FMEA action items.
• Field service/customer support	5. Install FRACAS system.	5. Structured FRACAS database and user input interface.
• Purchasing/supply management	6. Begin HALT Planning.	6. Detailed plan and schedule for HALT.
• Safety & regulation	7. Update lessons learned.	7. Updated lessons learned database, communicate to design team new issues and revise risk mitigation plan if needed.
	8. Review status and agree on each risk issue and mitigation plans. Risk mitigation meeting.	8. Risk mitigation meeting and agreement to proceed to next phase.

Note: DFM: Design For Manufacturing; DFR: Design For Reliability; DFT: Design For Test; DFS: Design For Service (and maintainability); FMEA: Failure Modes and Effects Analysis; FRACAS: Failure Reporting, Analysis and Corrective Action System; HALT: Highly Accelerated Life Test.

place in the design phase. The product development team has been relatively small up to this point and increases significantly to design the product, produce a working prototype, and create a documentation package. A well-designed and reliable product that is manufacturable, brings the added benefit of ensuring that the design team is released to work on the next project. If there are problems with the design, then engineering support will be needed to fix design problems. Pulling back key design resources to fix past problems negatively impacts resources needed to support future product designs. Some companies get around this by forming a sustaining engineering group. Their focus is to resolve design problems after a product has been released to manufacturing. The sustaining engineering group typically consists of less experienced designers and engineers to fix designs in which they were not involved. When there is a sustaining group to fix problems, it is vital that their findings make it back to the design team so that improvements are put in place to prevent the problem being repeated.

There is usually a cost associated with developing a more reliable design. Warranty cost savings will often justify these early-added costs. The capital expense, engineering resources, and time required to fix design problems can be significant. When sustaining engineering activity costs are added, this fuller picture will provide a better indication of what the actual cost is of an unreliable product.

Finally, the design phase is the last opportunity where the product design can be proactively improved. Once a product enters the design validation phase, the costs and schedule impact of design changes increase significantly. The problems found in product validation are more expensive to fix and can impact the product release date. Therefore, significant focus needs to be given in the product design phase to ensure that the product design is reliable.

There are eight reliability activities that take place in the product design phase (Table 15.1).

15.2   RELIABILITY ESTIMATES

As the program proceeds from concept to design, the materials required to produce the product are defined. Once the (BOM) is known, a reliability estimate can be performed. The reliability estimates are performed for each circuit board and subsystem that makes up the system. The estimate considers each item in the BOM and matches it to a reliability number. This reliability number is often expressed in FITs (Failures In Time per billion hours). Each BOM item's FIT number is added together to calculate the total FIT. The sum of all the individual FITs provides an estimate of the reliability for that board or subsystem. Reliability engineering is usually responsible for determining the reliability estimates for system, subsystems, and circuit boards. Estimating the reliability is more complicated when redundancy is involved or the components do not have a constant failure rate. The mathematics to deal with this is complex and beyond the scope of the book. Those who wish to gain a better understanding of this subject are referred to the bibliography section at the end of this chapter.

Reliability estimates provide insight into the areas where the product is likely to be least reliable. These estimates also point to those BOM items that are the major contributors to lowering the reliability estimate. There will not be sufficient resources (capital and personnel) to perform all the reliability tests on everything or to life test every component. By knowing where the reliability problems are likely to reside, the vital resources can be focused on the areas where the product is expected to be the least reliable. The reliability estimates are one technique to ensure that there is significant reliability focus around those areas of the product that are expected to be the least reliable.

The reliability estimates can also be used to determine if components that will not provide the reliability needed for the product are being selected. In these situations, you can evaluate alternative components and see the change they will have on the overall reliability. Reliability estimates may also help in determining the need to spend more for a more reliable component.

There is significant disagreement regarding the need to do reliability estimates and the value that they provide. First, reliability estimates can be significantly different from the reliability that you realize in the field. There are many factors that contribute to this uncertainty. Reliability estimates can vary significantly from actual observed reliability based on the component or device manufacturer, use conditions, applied derating, and the accuracy of the supplier supplied reliability data. Your ability to account for this variability will lead to a more realistic reliability estimate.

However, reliability estimates take a significant amount of time and resources to calculate. It is important to determine value is being received from the reliability estimates. It may be necessary to do a reliability estimate for contractual reasons or because customers have come to expect it. However, if this activity does not lead to reliability improvements, it should either be discontinued or the reason it is not providing value should be determined.

15.3   IMPLEMENTING RISK MITIGATION PLANS

The risk issues were first identified in the concept phase and high-level plans for mitigation were developed. The risk mitigation process continues in the product design phase. In this phase, detail plans for mitigating the risk are developed along with implementation. By the end of the design phase, all the details to build and test a working prototype are developed. As the system design architecture, bill of materials, schematics, and mechanical drawings are being developed, additional risk issues are identified and added to the risk mitigation plan. By the end of the design phase, all the risk issues should be identified with plans for mitigation. This occurs before the end of the design verification phase.

15.3.1 Mitigating Risk Issues Captured Reflecting Back

Using the 180-degree approach, risk mitigation issues were captured on retrospection and by looking forward. The strategies and plans for mitigating past problems are different from those looking forward. The strategies to mitigate known problems usually fall into one of four categories. The strategies to mitigate past problems are as follows:

1. Design out
2. Change use conditions
3. Fix part
4. Fix process.

Design out (or use an alternate part/supplier)

Parts that have had a history of being unreliable in design can often be designed out. There may be alternative design solutions that do not require using this component. It may even be possible to eliminate the part altogether. The purpose for the unreliable part may no longer be necessary. Designs are often recycled and modified for use in the next-generation design. Recycling designs avoids reinventing the wheel. However, with the passing of time, some of the circuitry may no longer be needed, either because its function is no longer needed, used, or required. The original designer may have moved on and the new designer copies it again. Design reviews and FMEAs can reveal these no longer needed situations. Another potential risk with recycling is part obsolescence. Designing in parts that are soon to be obsolete can cause undue delay in product development. On the other hand, recycling older designs will leverage the intellectual property of the organization. This is highly desirable as long as the designs are reliable.

An example of this is a design that used a three-phase motor for a cooling pump. When the pump motor is incorrectly wired, it may actually operate for a day or two, but soon the motor windings will burn up and the cooling system will fail. Replacing the motor is costly in dollars and lost production time. Over the years a special relay has been incorporated to circumvent this wiring error possibility. It is connected in such a way so as to allow the motor to run only when properly connected. This relay has helped to get new systems up and running without wiring mishaps. It is only needed the first time the system is activated. Once it is clear that the motor operates correctly, the relay has done its job. This solution has been used for many years and is a well-established safety item.

However, in this case the relay was high on the reliability Pareto. The contacts and magnetic windings were both prone to failure. Here, if the relay winding fails, the system will not work because the relay has to be energized before the correct windings are connected. So the relay winding failure may cause unnecessary shutdowns.

With today's new three-phase motor drives, power is connected directly to the drive electronics. The drive electronics is such that it doesn't matter how the three phases are wired because internal diodes redirect the currents correctly for all wiring cases. Miswiring is no longer an issue. Still, safety relays were used because of this recycling of intellectual property. Now because of the risk mitigation process, new designs that incorporate the motor drive electronics do not use a safety relay.

Over time, tried-and-true design solutions are incorporated without a thorough analysis as to its continued need. At times, as in the safety relay case, the old solution lowered the overall system's reliability.

Another reason to design out is because there are now better ways to achieve the same function, it may even require less parts. As electronics gets smaller and more integrated, it may be possible to replace the problematic electronics with new technology that has a proven track record for reliability.

Change use conditions

There may be situations in which a component is being used that has a history of being reliable but has been problematic in a particular design. Often, the cause is related to the use conditions. An example of this was observed with a semiconductor amplifier that the manufacturers specified as being highly reliable. The amplifier was being used in other designs without any problems. However, in this particular application, they failed at a high rate. The problem was related to an amplifier operating above its rated temperature specification. In this example, the part had a rated operating temperature of 85 °C, but exceeded this by 45 °C. The high operating temperature caused the device to fail early. Simply reducing the device temperature through improved cooling or more amplification stages running at lower power solved the problem. When problems like this are resolved, they should be added to a design review checklist or incorporated into the reliability design guidelines so they do not occur in a different design.

A significant portion of a product's reliability problems is design-related. Often the problem can be eliminated or the frequency of occurrence reduced to an acceptable level by decreasing the stresses that precipitated the problem. Changing the use conditions requires knowledge of the failure's root cause. Sometimes the problem is assumed to be because of a bad component even though the environment or the way it was used accelerated the failure. Getting to the root cause is a time consuming effort that may require testing and analysis, often by an outside service. However, if the root cause of the failure is not known, an inaccurate diagnosis will not resolve the problem.

Fix part

Some problems just need fixing. A good design concept that is poorly implanted needs to be fixed. This type of reliability problem is best fixed through a smarter design. The design fix can be mechanical, electrical, or both. It usually requires design modeling and later testing to prove out the design fix. The fix is often easy to implement in the design phase. The key was capturing it in risk mitigation so that resources can be allotted for redesign.

An example of this would be a part that has been unreliable and which has no better alternatives. This is common when working at the leading edge of technology. Once again, getting to the root cause of the problem is vital. This often requires working with the component manufacturer to determine if it is a design or process issue. Once the root cause is identified, suppliers will usually work with you to change the part. It is rare to encounter a problem with a component, with you being the only one in the industry experiencing it.

Fix process

Fixing the process is often required when the design is sound but either the manufacturing or test process is insufficient. It can be an assembly, rework, or test process that causes the problem. An example of this may be seen in component lead prepping. Lead prepping can damage a part, either through Electrostatic Discharge (ESD) exposure, residue left from a machining process, nicks in the lead, or stress cracks to the component body. It may be a moisture-sensitive part in which proper precautions are not being observed.

There can be problems with the test process as well. Examples are: ESD degradation from the testing process, lead damage through handling, and excessive moisture exposure.

15.3.2 Mitigating Risk Issues Captured Looking Forward

Looking forward, risk issues are captured that address what is new and unknown about the design. The risk mitigation activities looking forward answer questions regarding what the impact to product quality, reliability, and performance will be. The risk issues can be for a new technology, process, material, package, supplier, component, Printed Circuit Board (PCB), Application Specific Integrated Circuit (ASIC), Hybrid, design, manufacturing, or test process, as shown in Figure 15.2. The mitigation strategies resolve questions and concerns that can significantly impact the program if unresolved. Once answers to these questions are known, a decision can be made regarding the acceptability or fit for use in the design.

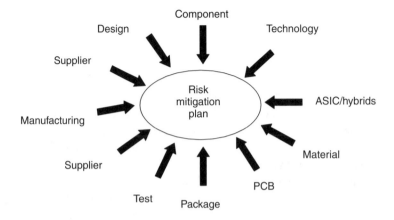

Figure 15.2   Looking forward to identify risk issues

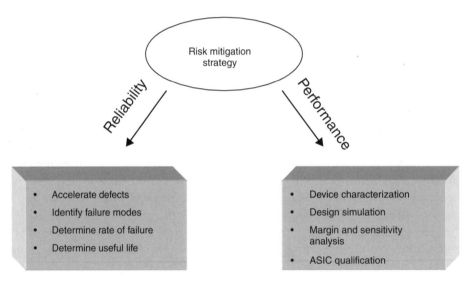

Figure 15.3   Risk mitigation strategies for reliability and performance

Getting answers to these questions requires testing. The testing to answer these questions will vary. We will consider two types, as shown in Figure 15.3. The testing addresses the reliability- and performance-related questions. The first type of testing (reliability testing) simulates various use operating conditions so device characterization, margining, and identifying the limits of use conditions can be found. This type of testing can also be time consuming and tedious. Many choose not to do device characterization. Those who don't will almost certainly regret it later on when design failures surface that could have been identified during this type of testing. The purpose of this type of test is to obtain an understanding of the device's behavior under a wide range of inputs, loads and use conditions, identify key operating parameters, and understand how design margin may be degraded by environment. Knowing this will help prevent designing in scenarios where the product may fail to operate. This is vital information for the designer but can only be learned through testing.

The other type of testing (performance testing) requires environmental stresses to stimulate defects. Environmental stresses are applied to cause (precipitate) a failure, determine a failure rate, determine the failure modes, and/or estimate useful life. Unfortunately, there is not a single test that will answer these questions. Instead, a test plan is developed on the basis of knowledge of what is expected to fail and why. Knowing this, accelerated test strategies are developed that will answer these questions. There is no universal test. This type of testing can be expensive and time consuming. If you are unsure what testing would work best, then using outside consultants is advisable. Environmental test facilities are a good source of knowledge about which tests are appropriate.

Often products are designed to have a useful life that is significantly longer than the amount of testing time that takes place in the design validation phase. It would be impractical to test the product for five years so you can verify that it will meet its specified useful life. To understand how a product or component will perform (over time) requires acceleration testing. Acceleration testing answers reliability risk issues regarding the following:

- How will it fail?
- When will it fail?
- What are the failures modes?
- When will it wear out?
- Are the failure modes accelerated by stress?

Knowing when the product fails in a more stressful environment and the root cause of the failure helps to answer these questions. Once this knowledge is known, improvements are made to remove the failure mode, reduce the frequency of occurrence, and/or extend the time it takes for the failure to occur. Many types of failure modes that are accelerated through environmental stress testing can be extrapolated to its frequency at a lower stress. The mathematics to determine the useful life based on an accelerated life test is best left to the reliability engineer. The results relate only to that particular failure mode. Most products have numerous failure modes. Therefore, several different accelerated life tests may be required to get a better understanding of the likely failure modes for a particular component, device, or design. The testing that is required to obtain this knowledge is called accelerated life testing.

Accelerated life testing

Accelerated life testing exposes a device to environmental stresses above what the device would normally experience, in order to shorten the time period required to make it fail. This type of stress testing quickly precipitates failures by compressing the time it takes to fail. The failures are the result of cumulative fatigue at exaggerated stress levels. For example, a paper clip that is bent open 90° and then back exaggerates the stress a paper clip will experience. If we repeat this process, we will accelerate the paper clip's useful life.

Accelerated life testing is performed because it is impractical to take a product designed for a ten-year useful life and then test it for ten years to verify that it conforms to specification. Accelerated life testing compresses the time it takes for failure to occur. The results from accelerated life testing are then used to verify that the product or device will survive for its designed service life.

The mathematics behind accelerated life testing is complex and best left to the reliability engineer to perform. However, the test process is easy to conceptualize. Either mathematically or empirically, an acceleration factor (acceleration rate) is determined for the accelerated life test performed and the intended use environment. By applying the acceleration rate, the useful life can be determined.

A mixed flowing gas test can be used to determine if a device will corrode in the field. This is an example of an empirically performed accelerated life test. In this test, the product is exposed to a mixed flowing gas in a controlled environment. The gas accelerates corrosion. There are different combinations of gas mixtures and concentration levels depending on the environment being simulated. Through empirical testing, it has been determined that, for example, two days of exposure relates to a year in your use environment. By knowing the product design life, it is easy to accelerate the corrosion that will occur in the device.

Suppose that a design requires a new connector that has many more contacts than anything you have used before. In addition, this connector is attached to the board using small solder balls that are smaller in size than previous designs. A possible risk issue for this connector is that it will not survive its required service life, and the reason believed is that the connector will have solder joint failures. An accelerated life test (in this case temperature cycling) will answer this question. To mitigate this risk issue, test boards are built with the connector soldered to the board using the standard manufacturing process. The boards are then placed in a temperature chamber and temperature cycled (i.e., 0 to 100 °C) to accelerate solder joint failures. By measuring the resistance at the solder joint, we can determine how many temperature cycles it will take to fail. The test is then repeated at a different temperature extreme (i.e., 0 to 130 °C) to determine the number of cycles to failure. After the two sets of tests are performed, a mathematical acceleration rate can be calculated. The acceleration rate is then used to determine if the connector will have solder joint failures during its service life.

The above two examples illustrate how accelerated testing can be used to evaluate performance over time. There are many different types of accelerated life tests. In addition, there are many different stress levels that can be applied for a particular test. This type of testing takes time and can be costly. However, it can be costly if you design a connector intended for a seven-year service life and it fails after three. First, *expect* every product in the field with this connector to fail around this time period found in the testing. Secondly, the failure will not be discovered for three years, thus there can be a significant amount of product affected. If the connector failure is a safety issue, then a recall is likely to be costly.

There are many different types of accelerated life test depending on the failures you wish to precipitate or evaluate (Table 15.3). Some common stresses used in accelerated life test are (Table 15.2):

Table 15.2   Common Accelerated Life Test Stresses

	Typical HALT Stresses
1	Temperature,
2	Vibration,
3	Mechanical shock,
4	Humidity,
5	Pressure,
6	Voltage,
7	Power cycling.

These stresses can be applied singly or in combination depending on the types of failures you wish to precipitate. The key to these tests is to keep the stress levels below the threshold where the physics of the material changes. If the stress that caused this failure is well above the physical limitations of the material(s) in test, the resultant failure(s) will not represent what can actually happen during customer use. In other words, if the stress temperature melts the plastic housing of a component, then you haven't accelerated its time to failure. Instead, you have identified and exceeded the physical limitations of the component. The information is not useful in determining the components useful life, only its upper use limits.

There are many different standard accelerated life tests used in industry (Table 15.2). There can be many variations in these tests. Some of the more common accelerated life tests are described below:

High Temperature Operating Life (HTOL)   This is an operational or biased test where the device is kept at an elevated temperature for an extended period of time. The primary purpose of this test is to accelerate failures that are the result of a chemical reaction. Examples are interdiffusion, oxidation, and Kirkendall

Table 15.3 Environmental Stress Tests

Accelerated stress test	Test conditions
High Temperature Operating Life (HTOL)	Temperatures vary as a function of the device in test.
Highly Accelerated Stress Test (HAST)	130 °C, 85% RH for 100 h
Autoclave	121 °C, 100% RH 103 kPa. Between 96 and 500 h
Temperature humidity bias (THB)	Typical: 85 °C & 85 RH Between 500 and 2,000 h
Temperature cycling	Varies 500–1,000 cycles typical −65–0 °C Low temp. 100–150 °C high temp.
Temperature storage	Varies 200 °C, 48 h 150–1,000 °C 125–2,000 °C 175–2,000 °C
Operating life	125–150°C 1,000–2,000 h
Thermal shock	500–1,000 cycles typical −65–125 °C 15 min dwells at each temperature
Random vibration	Varies greatly on the basis of user environment and product

voiding. Lubricant dry out can also be accelerated at elevated temperature. The acceleration rate can be modeled by the Arrhenius equation. The results of this test can be used to determine useful life at a lower temperature, that is, 65 °C.

Highly Accelerated Stress Test (HAST)   This can be a biased test where the device is kept at an elevated temperature in the presence of a controlled level of humidity for an extended period of time. The test is highly accelerated by using temperatures above the boiling point of water, 100 °C. The test is performed in a pressurized environment, where the pressure can be raised above one atmosphere. The primary purpose of this test is to accelerate failures that are temperature- and humidity-related. Humidity can cause material degradation, corrosion of metallization, degrade lead solderability, wire bond failures, bond pad delamination, intermetallic growth, and popcorning in plastic encapsulated components (moisture absorbed in package rapidly boils during assembly reflow and cracks the case).

Autoclave   This test is commonly referred to as a "pressure cooker" test. The device is placed in a pressurized chamber that has water stored in the bottom. The device is kept at an elevated temperature 121 °C, while suspended in saturated steam and is pressurized to 103 kPa (kilo Pascal). The concentrated steam is achieved by suspending the device at a minimum height of 1 cm above the water in the chamber. The test highly accelerates moisture penetration and galvanic corrosion.

Temperature Humidity Bias (THB)   This is an operational or biased test where the device is kept at an elevated temperature while in the presence of a controlled level of humidity for an extended period of time. The test is not performed in a pressure environment, the temperature is kept at 85 °C and the humidity level is held at 85% relative humidity. The primary purpose of this test is to accelerate failures that are temperature- and humidity-related. Humidity can cause material degradation, degrade lead solderability, and popcorning in plastic encapsulated components (moisture absorbed in package rapidly boils during assembly reflow and cracks component's body). The acceleration rate for the time to failure can be described using the Peck model. The results of this test can be used to determine useful life at a lower temperature and humidity, that is, 65 °C and 45% relative humidity.

Temperature cycling   This is a test where the device may or may not be powered during the test. The device is cycled to a low temperature extreme and dwelled typically for at least 10 to 15 min before it is transitioned to a high temperature and dwelled again for at least 10 to 15 min. The temperature transitions between high and low temperature extremes are continually repeated. The purpose of this test is to accelerate the effects of thermal expansion and contraction to see what fatigues. This is a common test to evaluate solder joint and interconnect reliability. The acceleration rate is modeled by the Coffin–Manson equation. The results of this test can be used to determine useful life knowing that the device will see less severe temperatures and at a lower frequency of occurrence (i.e., 65 °C when in operation and 25 °C when the device is turned off).

High temperature storage   This is a nonoperational test to accelerate temperature-related defects. The primary purpose of this test is to accelerate failures that are temperature- and humidity-related. These failures are the result of a chemical reaction that accelerates at elevated temperature. Examples are interdiffusion, oxidation, and Kirkendall voiding. Lubricant dry out can also be accelerated at elevated temperatures. The acceleration rate is modeled by the Arrhenius equation. The results of this test can be used to determine useful life at a lower temperature, that is, 65 °C.

Thermal shock This test is similar to thermal cycling except the time of transition between temperature set points is very short. The short transition time is achieved through a dual temperature chamber that can shuttle the product between the two chambers. Thermal shock can accelerate cracking and crazing of seals and encapsulated materials and hermetic package leaks.

Risk mitigation progress

It is important to track the progress that has been made to resolve risk issues. By the end of the design phase, the majority of the risk issues should be resolved. More importantly, the highest risk issues should be closed or nearing closure. One way to track the progress being made to mitigate risk issues is the risk mitigation growth curve, shown graphically in Figure 15.4. The risk mitigation curve illustrates the progress being made to mitigate risk. The slope of the curve indicates the rate at which new issues are being identified. When no new risk issues are surfacing, the curve will flatten out. Not all risk issues have the same severity. The risk issues are grouped into three categories, high, medium, and low. Each risk category is plotted separately so critical risk issues can be tracked separately to closure. The high-risk issues are the most significant and priority should be placed on these over lower-risk issues. The progress made against the medium and low-risk issues is also plotted in a reliability growth curve.

If high-risk issues are not being resolved, then proceeding to the next development phase may result in moving forward with your commitment for a

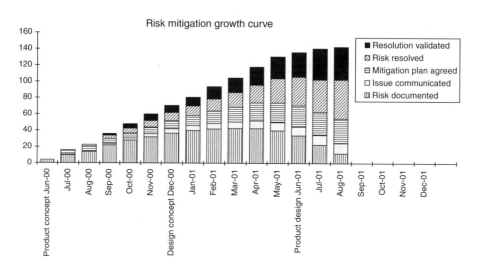

Figure 15.4 Risk growth curve shows the rate at which risk issues are identified and mitigated

program that is unlikely to succeed. The risk mitigation growth curve illustrates the progress made to mitigate risk issues.

By the end of the design phase, a significant portion of the program development resources has been spent. The resources required, both capital and manpower, to complete product development is significant. By the end of the design phase, the majority of the risk issues should have been resolved. The risk mitigation growth curves indicate the progress made and status of the effort to mitigate risk in the program.

15.4   DESIGN FOR RELIABILITY GUIDELINES (DFR)

Product development cycles are continually being compressed in development time. Shortening development cycles places further strains on the design team to develop a reliable product. Making matters worse, products today are designed to be smaller, lighter, faster, and cheaper. Each of these factors impacts product reliability. If the development cycles are too short, there may not be sufficient time to mitigate reliability issues and qualify new technologies, materials, suppliers, and designs before customer ship. As designers struggle to develop products in time to meet tight market windows, it is unlikely they will spend time addressing reliability issues. This problem can be resolved in part through reliability design guidelines that quickly aid designers with guidance on reliable issues. The reliability design guidelines are one of the few tools in the reliability toolbox that proactively improve product reliability. However, there are not many tools that can improve the reliability of a design before building and testing the first prototypes.

The reliability engineer can advise designers on ways to improve a design for reliability. They can identify components that are traditionally unreliable, suggest alternative design strategies that are more reliable, and evaluate the effect of derating and redundancy. However, unless there is a large reliability team, it is unlikely reliability engineering will be a part of every design decision. There are simply more designers making design decisions everyday than there are reliability engineers to review these decisions. In fact, it would be unrealistic to have a reliability review for every aspect of a design for reliability. Reliability should not be a policing function. It is more effective to develop DFR guidelines focused around past reliability problems and providing training on those guidelines along with updates to the guidelines. Training should be performed periodically to ensure that the design engineers know the DFR guidelines and discuss conflicts that they may have with the guidelines. It is vitally important that the design engineers understand how to apply the DFR guidelines since they are responsible for design decisions. The reliability engineer supports the design team explaining the need behind the guidelines and how they are applied. In situations in which the design guidelines cannot be followed, discussions take place between the designer and the reliability engineer for resolution.

Recall reliability is the responsibility of the designer, not the reliability engineer. The designer is ultimately responsible for the product design, its reliability, manufacturability, and testability. This sounds like an unrealistic requirement. Designers generally do not know what to do to improve the reliability of a design. Having DFR guidelines allows the designer to make the right decisions before the product is ever tested. The reliability engineers are responsible for developing these guidelines and to provide guidance on their implementation. Reliability engineers are also responsible for working with the designer when a reliability requirement cannot be met.

The design guidelines are defined in black and white, but there will be gray areas in their interpretation. Unfortunately, not every application is black and white. These shades of gray represent potential risk areas and sometimes require testing or outside expert opinion for resolution. As the gray areas are resolved, the DFR guidelines are revised to reflect changes. The guidelines will probably not cover every aspect of the design, especially those associated with leading edge technologies. These issues are better captured in a risk mitigation plan with the results incorporated into the design guidelines. The guidelines are to be continually revised to reflect continuous improvements, and reflect present technology and new reliability issues that surface in the existing product.

Applying reliability guidelines involves evaluating the trade-offs that affect product failure rate, repair cost, safety, image, profit, and time to market. The design decisions for increased reliability can have a negative impact on the product cost, ease of manufacturing, and design complexity. It is best to avoid over design or designing a product that no one is willing to pay for. This requires a marketing understanding of the reliability requirements, design trade-offs, liabilities, and warranty cost and consumer impact. The impact from selling an unreliable product may not be known for years. However, expect a trend of reduced market share as word spreads out about customer dissatisfaction. The cost of an unreliable product also manifests itself in rework, scrap material, and expensive recalls identified late in the product life cycle.

There is an additional cost factor associated with lost repeat business from a dissatisfied customer. Today's consumer is much more knowledgeable about product reliability. With the advent of the Internet, dissatisfied customers have a much greater impact on future sales through chat rooms and customer product reviews that are now found at popular sites. This may be the greatest threat. Today's consumer is computer smart and can easily research a product's reliability history. Unfortunately, most of the product reviews on these web pages are the result of dissatisfied customers. To survive, the products developed and produced must be reliable. The best way to achieve this and to meet the demanding time-to-market requirements is by incorporating DFR guidelines in a concurrent engineering effort.

Most companies today have DFM and DFT guidelines as part of the product development process. The benefits from DFM and DFT are well understood. Design teams understand how DFM and DFT reduce product development time and reduce the number of engineering changes at product release. The DFM and DFT guidelines are applied in the design phase and usually incorporate a checklist to verify that past mistakes are not repeated. These DFM and DFT guidelines have been developed from lessons learned over time. These lessons are then communicated through a set of guidelines and become part of the checklist required for final design approval and sign-off.

The same techniques and processes that are used in DFM and DFT apply to reliability design guidelines. The process for applying design guidelines is already in place; the problem for most companies is that they do not have reliability design guidelines and do not know how to create them. When it comes to reliability design guidelines, it seems we are back in the Stone Age where the design is thrown over the fence to manufacturing. Then it becomes the responsibility of the quality and manufacturing team or sustaining engineers groups.

So how do you establish DFR guidelines? The DFR guideline format is the same one used for the DFM and DFT guidelines. In fact, you may find that reliability issues are also covered in DFM and DFT guidelines. This is fine as long as the guidelines don't conflict. If there are debates over where the guidelines belong, refer back to the basic definitions for each charter. DFM guidelines focus on issues affecting manufacturability, cost, quality, product ramp, and rework. DFT guidelines focus around the testability, test access, fixturing, and safely operating the Device Under Test (DUT). DFR guidelines focus around product quality over time. If DFR guidelines do not exist, chances are that some of the reliability issues are being covered in the DFM and DFT design guidelines.

The DFR guidelines can take several years to develop, so it is unreasonable for designers to wait that long for guidance on how to design for reliability. The guidelines must take into account the different users. The guidelines are developed by a team and address the needs of all the design groups. The guidelines are consensus driven. If no guidelines exist, start by creating a Pareto of reliability issues that need guidelines. This is usually derived from field failure data, customer complaints, and manufacturing ramp/test issues. Once created, it then becomes a task of creating, training, and implementing each guideline as it is developed.

The DFR guidelines call out technologies, components, and packages that should be avoided or not used. Be sure while identifying what not to use, that you offer recommendations as to what is the best alternative.

Reliability design guidelines should be organized for easy access of information. Putting the guidelines into a searchable, electronic database is highly desirable. Over time the guidelines will grow into a significant volume of knowledge (which should be treated as corporate intellectual property). A suggested reliability design guidelines table of contents would be as follows:

1.0   Introduction – the need for reliability
2.0   Component reliability guidelines
3.0   Mechanical reliability guidelines
4.0   System reliability guidelines
5.0   Thermal reliability guidelines
6.0   Material reliability guidelines
7.0   System power reliability guidelines
8.0   Reliability safety guidelines.

Each guideline should be defined in a single page if possible. The guideline should address a single thought. In other words, if you were developing reliability design guidelines for capacitors, dedicating single pages for each type of capacitor (i.e., electrolytic, ceramic, tantalum) will make it easier for the end user to apply. Each guideline should address the following:

What is the reliability design requirement?
What is the impact if not followed or the benefit if followed?
What detail is required to properly apply guidelines?

The above three questions should be answered for each reliability design requirement. An example of this is shown in Figure 15.5.

Unfortunately, DFR guidelines cannot be purchased. Most businesses develop their own set of guidelines tailored to their particular business.

15.4.1   Derating Guidelines

Derating guidelines are vital to any product development program and should be a part of the DFR guidelines. If there are no derating guidelines in place, chances are good that some of the customer failures are the result of stress levels that exceeded the component specifications. In addition, having sufficient derating in the design will result in increased design margin. Designs with sufficient design margin have lower test Parts Per Million (PPM) failure rates and higher first-pass yields in production.

It is not necessary to develop derating guidelines from scratch. There are several derating guidelines in the industry that can be purchased. The guidelines cover a broad range of users and so they may not be usable by engineers in this form. However, it is a simple task to tailor derating guidelines for your specific

Requirements/options:	Reliability impact/benefit:
1. Temperature stress (Rule #1): For every 10 °C increase in temperature the useful life decreases by a factor of 2. 2. Ripple current stress (Rule #2): Stay below 50% of the maximum ripple allowed. 3. Voltage stress (Rule #3): For voltage stress derating above 67%, the expected life increases by the fifth power of the rated voltage (V_r)/applied voltage (V_a).	Derating greatly improves the life of the electrolytic capacitor in addition to ensuring greater protection from spikes that can cause shorted capacitors.

Detail:

The expected life of a capacitor is described as the maximum expected life (Condition a) at rated temperature times the acceleration factors: temperature (T), voltage (V) and ripple current (I).

$$L_b = L_a \times A_t \times A_r \times A_v$$

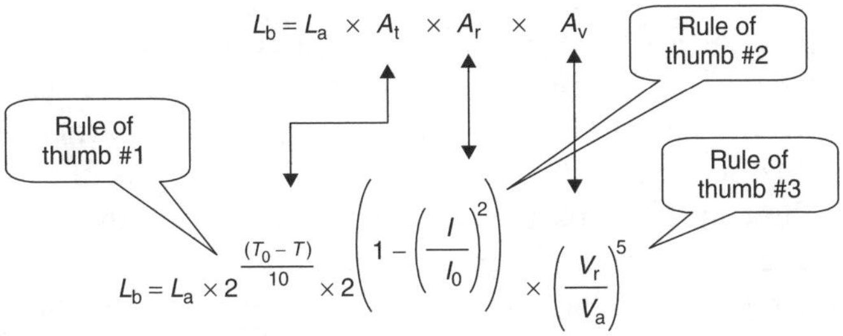

Rule of thumb #1

Rule of thumb #2

Rule of thumb #3

$$L_b = L_a \times 2^{\frac{(T_0 - T)}{10}} \times 2^{\left(1 - \left(\frac{I}{I_0}\right)^2\right)} \times \left(\frac{V_r}{V_a}\right)^5$$

L_a = Lifetime under condition a,
L_b = Lifetime under condition b,

A_t = Temperature acceleration,
A_v = Voltage acceleration,
A_r = Ripple current acceleration.

Figure 15.5 DFR guideline for electrolytic capacitor usage. Courtesy of Teradyne, Inc.

application and user environment. Derating guidelines are available from the Reliability Analysis Center. The ordering information is as follows:

Electronic Derating for Optimum Performance
Reliability Analysis Center
201 Mill Street
Rome, NY 13440-6916
http://rac.iitri.org/

Once the guidelines are in place, they can become part of the design review process. Derating should also be part of a design check off to verify compliance to the guideline prior to design review.

15.5   DESIGN FMEA

The most powerful reliability tool in the design phase prior to building and testing any product is the FMEA tool. The design FMEA needs to be performed prior to signing off designs for procurement and build. The design FMEA activity takes place before any prototypes are ever built. If you've never done FMEAs in your organization, this is one area where resistance is almost guaranteed. For some reason, designers have a hard time embracing the concept of a design FMEA. They will often point to the many design checks that are incorporated into the design process to eliminate errors as good enough, with "This is how we've always done it." Typically, design reviews to catch problems include a checklist based on common mistakes made, peer reviews, automated simulation programs, and automated design check programs to verify compliance to design guidelines. These processes are valuable and necessary to the design process. However, they have limitations in the types of problems they can identify. Because the FMEA is a concurrent effort, potential reliability and safety issues can be identified and fixed where it will have the least impact on the program.

In Chapter 7, the FMEA process is discussed in detail. We recommend that the material presented in Chapter 7 be used to develop an FMEA training program. Prior to a design FMEA, it is imperative that all participants have been trained in the process. Performing design FMEAs for the first time in an organization is difficult. If the participants are not trained in the process, then the FMEA meeting will be highly unproductive. Untrained participants have a tendency to steer the team off on tangents and are more likely to challenge the process. It is better to delay the FMEA design review by a week so that everyone is trained than to train as you go through the design FMEA. The training doesn't have to be long, but everyone needs to be familiar with the process.

FMEAs should be performed for all significant sections of the design. This includes the total system and subsystems including Printed Circuit Board Assemblies (PCBAs).

The best place to perform the design FMEA is after the design is completed but before the design goes through final design review. Design changes usually result from the FMEA, and so there is no point in doing the final design review until the FMEA items have been closed out. The benefits from a design FMEA include:

- discovery of design errors,
- identification of system failures due to interconnects,
- identification of failure effects from grounding problems,
- identification of failure effects if voltages sequence at different times,
- analysis of impact when high-risk reliability components fail (i.e., what is the likely failure effect when a tantalum capacitor fails short?),

- identification of safety, regulatory, or compliance issues,
- identification of failure effects due to software errors,
- test comprehensiveness.

The output from the design FMEA is a list of design issues that require corrective action. The corrective action list is order ranked on the basis of the severity of each issue. After the team has completed the FMEA spreadsheet and the corrective action list has been generated, the next step is to decide what issues will be resolved and who will do it. There usually is confusion at this point over which issues should be resolved. Obviously, all safety, regulatory, and compliance issues need to be addressed. There unfortunately is no standard rule for deciding which of the nonsafety-related issues should be resolved. Factors like available resources and time available to fix issues need to be considered. Some companies use the 80/20 rule where the top 20% (corrective action issues) represents 80% of the potential problems. Once it is agreed upon which issues will be fixed, the next challenge is tracking these issues to closure. Often, issues are identified as needing to be fixed but because there is no follow-up, they remain unresolved. A simple solution is to generate a single form for tracking FMEA issues to closure.

15.6 INSTALLING A FAILURE REPORTING ANALYSIS AND CORRECTIVE ACTION SYSTEM

Failure Reporting Analysis and Corrective Action System (FRACAS) is a closed loop feedback system used to collect and record data, analyze trends, and track problems to root cause and corrective action(s) for both hardware and software problems. A FRACAS system provides a cradle-to-grave solution for problem resolution. FRACAS is used to verify containment and resolution of failures. A good FRACAS system identifies reliability problems when they surface, tracks the progress made in identifying root cause and corrective action. Finally, FRACAS is used to track problems to closure and without it, the impact to the bottom line can be significant and problem identification/resolution may be nothing more than guesswork.

The FRACAS system is installed in the design phase. The installation can be as simple as structuring the new product into the FRACAS database and ensuring that appropriate data entry fields are in place. FRACAS is first operated during prototype, where design bugs are entered after they are identified. The designers are responsible for entering this data; therefore they should be adequately trained and should be familiar with the FRACAS software and database. They should be able to easily access the database and it should have sufficient capacity to manage the volume of activity anticipated. If the FRACAS system is new, then sufficient time should be allowed to debug the software before beginning construction of the prototype. If the FRACAS system is buggy, clumsy, or difficult to use when prototyping begins, it is likely that design problems will not be entered into the database. Instead, they will be recorded in notebooks, scrap paper, and personal computers in which they may be misplaced or lost and never tracked to closure.

Implementation of the FRACAS system will require the following:

1. Identification of the key product parameters that will be used to sort the information (i.e., date, manufacturer, part number, quantity, where used on, etc.). This is a much longer list but existing failure report forms can be used for this source.

2. Deciding if the FRACAS system will be manual (paper system) or if a computerized method will be chosen. This is not a trivial task, especially if it is computerized. The FRACAS system will take into consideration everything that is nonconforming or unacceptable from:

   (a) engineering development data,

   (b) FMEA recommendations,

   (c) HALT findings,

   (d) Highly Accelerated Stress Screens (HASS)/Highly Accelerated Stress Audit (HASA) findings,

   (e) incoming material Inspection nonconformance,

   (f) in-process manufacturing failure reports,

   (g) field failure reports,

   (h) customer feedback.

3. The identification of the personnel who will sit on and the one who will lead the Failure Reporting Board (FRB). The Quality Manager often leads the FRB. The FRB lead must have the authority to drive all the issues to closure. The board will consist of personnel from manufacturing, purchasing, design engineering (sustaining engineering), marketing, product management, and perhaps others.

15.7 HALT PLANNING

Planning is a major part of HALT testing and can consume more time than the test itself. There are many issues that need to be worked out before HALT begins. First, there needs to be agreement regarding what will undergo HALT. After deciding the assemblies for HALT, there needs to be consensus on the number of assemblies that will be tested. There should be at least three to five assemblies for HALT with an additional unit used for debugging only (a "gold" unit). HALT is a destructive test. After the test, the assemblies cannot be repaired and sold because a considerable amount of product life has been removed. If the assemblies are expensive, then debate is likely regarding the number of units that are destructively tested. Avoid testing only a single unit. This is impractical if you are doing the test at a test facility. The problem with having only one unit to test is that there will be significant downtime after each failure. The time it takes to troubleshoot and fix a failure can be significant. If there are multiple units, then testing can proceed with the next unit as you troubleshoot and fix the failed assembly. HALT planning flow is illustrated graphically in Figure 15.6.

The design team and its management must buy in. Management's commitment of resources to support the HALT effort communicates the commitment that reliability activities will be performed to improve product design. After it is agreed which assemblies will be HALT tested, the next step is to form HALT teams for each assembly. The teams are cross-functional and consist of members from software, test, manufacturing, design engineering, and reliability. The purpose of the cross-functional teams is to work out all issues related to supporting

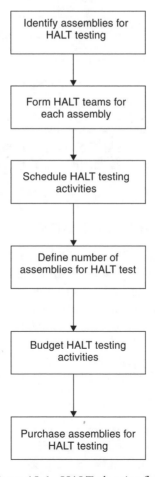

Figure 15.6 HALT planning flow

			HALT Planning Meetings held at_____AM/PM	
Latest rev.			= Need to discuss this week *(HALT Planner it to highlight AI# for discussion at next HALT Planning Meeting)*	
mm/dd/yyyy			= Action complete (HALT leader is to *darken the Done Date when action is completed)*	
Action owner	Done date	AI#	Activities	Notes
		1	HALT week date set	
		2	Lab is available for HALT week	Contact name, address & phone
		3	HALT Team identified	
		3.1	Designer	Name & phone
		3.2	Software	Name & phone
		3.3	Test	Name & phone
		3.4	Reliability	Name & phone
		3.5	Chamber technician	Name & phone
		3.6	Repair facilities	Contact name, address & phone
		4	Liquid nitrogen is available for HALT week	Order tank refill if needed
		5	Assemblies (DUTs) are available for HALT week	
		6	Extra interface unit(s) needed for HALT	Yes–no
		7	Extra interface unit(s) is available for HALT	
		8	Cables to connect from instruments to DUT are available	Are spare cables needed?
		9	Power supplies are available for HALT	
		10	Power supply cables available for HALT	Are spare cables needed?
		11	Mechanical fixturing is available for HALT	
		12	Mechanical fixturing verified that it works with DUT	
		13	Test instrumentation for HALT test is identified	
		14	Test instrumentation is available for HALT	
		15	Make list of things to bring to HALT Lab	
		16	Make list of things to ship to HALT Lab	

Figure 15.7   HALT planning check list

the HALT effort prior to the test. Use a checklist to ensure that all issues are addressed. An example of the HALT checklist is shown in Figure 15.7. If you plan to outsource HALT to a test facility, this step will help in managing test cost and test time.

15.8   HALT TEST DEVELOPMENT

In the HALT planning phase, the assemblies were identified for HALT testing along with the number of assemblies to be tested. Teams were then formed for each assembly to support the HALT activities. The HALT test development can begin after the planning is in place. The goal of test development is to have everything ready to support the HALT effort before testing begins (Figure 15.8). In HALT, development teams are formed for each assembly and will vary depending on the specific skills needed. The first activity of the team is to define the HALT stress tests that will be performed. Defining the HALT test starts with identifying which stresses will be exerted on the assembly to reveal reliability concerns.

After the stresses are identified for the each assembly (Table 15.2), upper and lower stress limits may need to be identified. There may not be an upper or lower limit that is known. Limits are defined when the component changes physical states due to a known stress level. An example of an upper temperature

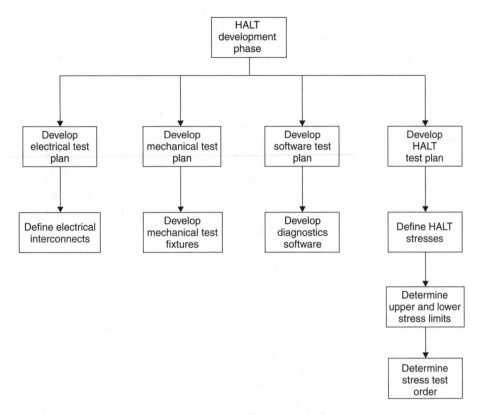

Figure 15.8   HALT development phase

limit in an assembly is a temperature that causes a connector housing to melt. There's no reason to stress an assembly beyond a known physical limitation. The failures identified when the upper limit is exceeded do not relate to real field failures.

After the HALT test plan is defined, the remainder of the HALT test development activities can precede. The three areas of activity in the HALT planning are

- mechanical fixturing,
- electrical test plan and execution,
- software test plan and execution.

The mechanical test plan includes defining how the assembly will be mechanically fixtured in the HALT chamber. Mechanical fixturing should optimize the energy transfer from the vibration chamber into the device under test. Mechanical fixturing should not induce resonances into the assembly. The fixturing should be as light as possible and mechanically strong. There are several companies that make universal mechanical fixturing. A list of these companies can be found in Appendix A. After the mechanical fixturing is developed, it is a good idea to test the fixturing by attaching accelerometers to the device under test. Place an accelerometer on the vibration table and several on the DUT to verify that the mechanical energy transfer is efficient. On the screw type fasteners, use mechanical locking devices like split lock washers to ensure the DUT is securely attached to the vibration chamber.

Developing the electrical test plan is usually more complicated. Ideally, it should be the same test plan that is being developed for the assembly in manufacturing. However, it is not unusual for the manufacturing test plan to be incomplete at the time of HALT testing. The test plan for manufacturing may also not transfer well to a HALT test. In-circuit bed of nails fixturing designed for manufacturing test will not perform well in a HALT chamber. The test development team may need to develop special fixturing for the HALT test. The electrical test plan must include how to power the DUT and what I/O signals will be connected to the assembly during test. The strategy for HALT testing is to have only the assembly under test in the HALT chamber and all external supplies, support logic, loading, and test I/Os external to the chamber. This may not always be easy.

Developing the software test plan is usually more straightforward. The software that is developed for testing the product in manufacturing can usually be used for the HALT test. It is important to identify the software that will be required for the HALT test and to make sure it is ready at the time of the test. The software that will be used for HALT needs to be checked out before HALT testing.

Before HALT begins, the following questions need to be answered:

1. What assemblies will be tested?
2. What assemblies can be tested?
3. What must be omitted?
   For each particular assembly:
4. Quantity of each assembly for test
5. What testing is required to verify proper operation?
6. What testing can be performed to verify proper operation in the HALT chamber?
7. What software is required to do the HALT testing?
8. What hardware is required to do the HALT testing?
9. Who will build the boards (use production process and tooling)?
10. Who will debug the board?
11. What mechanical test fixture(s) is required to do the HALT testing?
12. Any special electrical test fixtures required to do the HALT testing?

13. What cabling and interconnect is required to do the HALT testing?
14. Any special cooling plate required for HALT testing?
15. Any special power requirements needed to do the HALT testing?
16. Any special test equipment required to do the HALT testing?

Logistics and scheduling issues:

1. What is the material cost for the reliability assemblies that will be tested?
2. What is the engineering development time and resources required?
3. What is the software engineering development time and resources required?
4. What is the mechanical fixturing development time and resources required?
5. What is the engineering test development time and resources required?
6. What is the manufacturing development time and resources required?

15.9   RISK MITIGATION MEETING

By the end of the design phase, a significant portion of the product's development resources has been expended. A risk mitigation meeting is scheduled as the project nears completion of the design phase. The meeting should focus on the progress that has been made to mitigate the most severe risk issues. The progress made to mitigate the most severe risks before first prototype is a strong barometer of how the program is being managed. If the rate of new risk issues (risk mitigation slope) has not flattened out, then there is a good chance that the design is still in a state of flux. If the rate of closure on risk issues is not increasing, it can be a sign that there is a lack of commitment to resolve key issues. If the most significant risk issues are not addressed and mitigated until late in the program, then significant redesign and program setbacks are possible. In addition, if satisfactory progress has not been made on the most severe risk issues, alternative solutions must be initiated.

The risk mitigation meeting should focus on these issues. The functional groups meet periodically to review progress and strategies for closure on each risk issue. The purpose of the meeting is to determine if adequate progress has been made and whether the program should proceed into the design validation phase. The risk mitigation meeting reports to senior management on the progress made by the individual groups to mitigate risk since the last development phase. There is no need to expend significant capital for prototypes if the project is not likely to succeed.

References

FMEA

1. *Recommended Failure Modes and Affects Analysis (FMEA) Practices for Non-Automobile Applications*, SAE (2001).

2. M. Krasich, *Use of Fault Tree Analysis for Evaluation of System Reliability Improvements in Design Phase*, 2000 Proceedings Annual Reliability and Maintainability Symposium (2000).
3. K. Onodera, *Effective Techniques of FMEA at Each Life-Cycle Stage*, 1997 Proceedings Annual Reliability and Maintainability Symposium, IEEE (2000).
4. S. Bednarz, Douglas Marriot, *Efficient Analysis for FMEA*, 1998 Proceedings Annual Reliability and Maintainability Symposium (1998).
5. M. Kennedy, *Failure Modes and Effects Analysis (FMEA) of Flip-Chip Devices Attached to Printed Wiring Boards (PWB)*, IEEE/CPMT International Manufacturing Technology Symposium, IEEE (1998).
6. R. Whitcomb, M. Riox, *Failure Modes and Effects Analysis (FMEA) System Development in a Semiconductor Manufacturing Environment*, IEEE/SEMI Advanced Semiconductor Manufacturing Conference, IEEE (1994).
7. D. J. Russomanno, R. D. Bonnell, J. B. Bowles, *Functional Reasoning in a Failure Modes and Effects Analysis (FMEA) Expert System*, 1993 Proceedings Annual Reliability and Maintainability Symposium, IEEE (1993).
8. S. Prasad, *Improving Manufacturing Reliability in IC Package Assembly Using the FMEA Technique*, IEEE Transactions of Components, Hybrids and Manufacturing Technology, **14**(3), 452–456 (1991).

HALT

1. General Motors Worldwide Engineering Standards, *Highly Accelerated Life Testing*, GM (2002).
2. G. K. Hobbs, *Accelerated Reliability Engineering*, John Wiley & Sons (2000).
3. H. W. McLean, *HALT, HASS & HASA Explained: Accelerated Reliability Techniques*, American Society for Quality (May, 2000).
4. J. Strock, Product Testing in the Fast Lane, *Evaluation Engineering* (March, 2000).
5. N. Doertenbach, *High Accelerated Life Testing – Testing with a Different Purpose*, IEST, 2000 Proceedings (February, 2000).
6. D. Rahe, *The HASS Development Process*, 2000 Proceedings Annual Reliability and Maintainability Symposium, IEEE (2000).
7. D. Rahe, *The HASS Development Process*, ITC International Test Conference, IEEE (1999).
8. M. Silverman, *Summary of HALT and HASS Results at an Accelerated Reliability Test Center*, Qualmark Corporation, Santa Clara, CA, 1998 Proceedings Annual Reliability and Maintainability Symposium, IEEE (1998).
9. M. Silverman, *HASS Development Method: Screen Development, Change Schedule, and Re-Prove Schedule*, 1998 Proceedings Annual Reliability and Maintainability Symposium, IEEE (1998).
10. R. H. Gusciaoa, *The Use of Halt to Improve Computer Reliability for Point of Sale Equipment*, 1998 Proceedings Annual Reliability and Maintainability Symposium, IEEE (1998).
11. J. A. Anderson, M. N. Polkinghome, *Application of HALT and HASS Techniques in an Advanced Factory Environment*, 5th International Conference on Factory 2000 (April, 1997).

12. M. L. Morelli, *Effectiveness of HALT and HASS*, Hobbs Engineering Symposium, Otis Elevator Company (1996).
13. C. Ascarrunz, *HALT: Bridging the Gap Between Theory and Practice*, International test Conference 1994, IEEE (1994).
14. R. Confer, J. Canner, T. Trostle, S. Kurz, *Use of Highly Accelerated Life Test Halt to Determine Reliability of Multilayer Ceramic Capacitors*, IEEE (1991).
15. P. E. Joseph Capitano, Explaining Accelerated Aging, *Evaluation Engineering*, p. 46 (May, 1998).
16. E. R. Hnatek, Let HALT Improve Your Product, *Evaluation Engineering*.
17. G. K. Hobbs, What HALT and HASS Can Do for Your Products, *Evaluation Engineering*.
18. G. K. Hobbs, What HALT and HASS Can Do for Your Products, *Hobbs Engineering, Evaluation Engineering*, Qualmark Corporation, p. 138 (November, 1997).
19. E. O. Minor, *Quality Maturity Earlier for the Boeing 777 Avionics*, The Boeing Company.
20. M. A. Silverman, *HALT and HASS on the Voicememo II™*, Qualmark Corporation.
21. M. Silverman, *Summary of HALT and HASS Results at an Accelerated Reliability Test Center*, Qualmark Corporation, Santa Clara, CA.
22. M. Silverman, *Why HALT Cannot Produce a Meaningful MTBF Number and Why This Should Not be a Concern*, Qualmark Corporation, ARTC Division, Santa Clara, CA.
23. W. Tustin, K. Gray, Don't Let the Cost of HALT Stop You, *Evaluation Engineering*, pp. 36–44.

16

Design Validation Phase

In the previous phase, the schematics, bill of materials, and outline drawings needed to design the product were developed. In addition, functional prototypes exist. Now, in the design validation phase, the functional prototypes are tested to verify that the design conforms to specification. This is the final opportunity to identify design, quality, reliability, manufacturing, test, and supplier issues before the design is released for production. Identifying all the design-related problems takes a cohesive effort between manufacturing engineering, test engineering, reliability, and design engineering to fully evaluate the design. At this point in time all of these functional groups are working on the program in the design validation phase. Each has different concerns regarding the reliability of the product. Everyone is diligently working to resolve any remaining risk issues prior to production release. Manufacturing is validating special tooling and assembly processes for rampability. Test engineering is checking out test hardware, software, and test fixtures. Reliability is stressing the product to understand how it will fail. Engineering is testing prototypes to validate that the design meets the concept requirements with margin. The majority of the design problems (bugs) are identified in this phase.

This is engineering's last opportunity to identify and fix design-related problems before shipping the product to the customer. Once the product is released to production, the design team will be redirected towards the next platform or a derivative product. If design-related problems surface later in production, resolution of the problem usually is not the responsibility of the original design engineer. Instead, sustaining engineering will support this activity. This group probably will not have the technical experience and knowledge of the original design team. That is why it is so important to identify and fix design problems in the validation phase. The activities that take place in the validation phase are shown in Table 16.1.

Improving Product Reliability: Strategies and Implementation. Mark A. Levin and Ted T. Kalal
© 2003 John Wiley & Sons, Ltd ISBN: 0-470-85449-9

Table 16.1 Reliability Activities in the Validation Phase

	Validate design phase	
Participants	Reliability activities	Deliverables
• Reliability engineering	1. Design & performance validation.	1. Product performance specifications are validated, and any limitations noted.
• Marketing	2. HALT working prototypes. Failures traced to root cause and corrected in design. Product undergoes a Final HALT to verify fixes.	2. HALT failures, stress levels, and root cause document in report. Corrective action plan to remove failure modes. Final HALT report verifies fix.
• Design engineering (electrical, mechanical, software, thermal, etc.)	3. HASS effectiveness is validated using a Proof Of Screen (POS) test.	3. POS verifies effectiveness of HASS protocol.
• Manufacturing engineer	4. Operate FRACAS. All failures during prototype are entered into FRACAS database and tracked to closure.	4. FRACAS report.
• Test engineering	5. FMEA is performed (on any significant design changes only) & process FMEA.	5. Completed FMEA spreadsheet and closure on corrective action items.
• Field service/customer support	6. Closure on all risk issues. Review status and agree on each risk issue and mitigation plans. Risk mitigation meeting.	6. Risk mitigation meeting and agreement to proceed to next phase. Risk issues need to be mitigated before production phase.
• Purchasing/supply management		
• Safety & regulation		

Note: FMEA: Failure Modes and Effects Analysis; FRACAS: Failure Reporting, Analysis and Corrective Action System; HALT: Highly Accelerated Life Test; HASS: Highly Accelerated Stress Screens; POS: Proof Of Screen.

16.1 DESIGN VALIDATION

At the end of the design phase, purchasing obtains material for prototype testing and evaluation. Boards are then built using the standard manufacturing process. Design engineers should not build the prototypes because this is an opportunity for manufacturing and test to spot problems early. Once built, engineering begins the process of validating the design's conformance to the requirements and specifications laid out in the concept phase. To fully validate the design takes time and patience. At this stage, unfortunately, many programs find themselves behind schedule and over budget. There is a natural tendency to shortcut the design validation process so that production can begin. Shipping without completing design validation will certainly result in significant retrofitting, Engineering Change Order (ECO) activity, and higher failure rates. There is significant risk in releasing a product without knowing how it will perform under various customer environments and use conditions.

Design validation is the process of testing the design to learn how it will perform under various loads, inputs, environments, and use conditions. In essence, you are characterizing the performance capabilities and identifying the limitations in the design. Design validation testing also involves accelerated stress testing so that potential field failures can be identified. Once this information is known, the design can be refined to increase performance and enhance reliability. In addition, the product performance capabilities can be defined on the basis of what the design is capable of achieving. If certain aspects of the design specifications cannot be met, then reducing the product specification prior to release is possible. The design verification process improves both the reliability and performance of the design, as shown in Figure 16.1.

Design validation testing begins by testing the device's performance under ambient or nominal conditions. The device is then tested at the upper and lower specified temperature operating extremes to verify that it can operate safely and to specification. Then the product is tested at temperatures in excess of the specification to determine how much design margin there is. Design margin is important because there is a relationship between design margin and first-pass

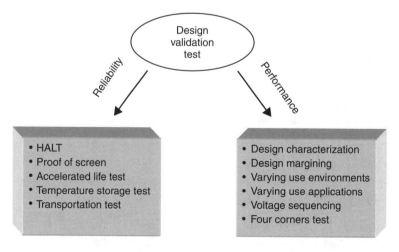

Figure 16.1 Reliability activities in the validation phase

yield in manufacturing. Designs that have sufficient design margin also have high first-pass test yields. Conversely, if there is no design margin, the product is likely to have a higher failure rate in test. Design margin can compensate for component variability.

16.2   USING HALT TO PRECIPITATE FAILURES

HALT testing is performed on circuit boards, subassemblies, and at the system or product level. System level HALT can be difficult if the system is physically large in size. Three problems exist. One, most HALT chambers do not accommodate large systems. Two, it can be difficult to get sufficient vibrational energy into a large system to precipitate a failure. And three, temperature changes in the product in the chamber can be prohibitively long for efficient HALT. The only place where HALT testing is not performed is at the component level. Accelerated life testing, as defined in Chapter 15, is the best way to accelerate failures, identify failure modes, and determine reliability at the component level.

At some point in this phase material will need to be purchased and assemblies built for HALT. For expensive assemblies, it is best to hold off buying the costly items for HALT until there is assurance that the design works. This isn't always possible if the items have long lead times, minimum purchase quantities, or high Non Recurring Engineering (NRE) charges. If all the material is bought for HALT but the prototypes do not work, some of the material may well end up being scraped. The risk is greater on a new platform product than a derivative. It is also not recommended to perform HALT on a board that has had a massive amount of engineering fixes, jumper wires, and glued components. The problem with excessive amount of rework is that issues can surface in HALT due to the quality of the rework and not the quality of the design and manufacturing process. Circuit boards that have excessive amounts of rework should have the board artwork revised to incorporate the design changes. It is best to consult design engineering after the prototypes are built to get an early indication of the functionality of the prototype.

HALT should be performed on assemblies that are built using the same bill of material as the final product. Using material from a different supplier for prototype and HALT may identify problems with a prototype part or process that is not in the final product. In addition, some problems will go unidentified because problems with a component or manufacturing process cannot surface if it is not part of the prototype process. Some examples of parts that are different in prototype from the final product are as follows:

- Machined parts. They will respond differently under stress from a cast part
- Hand soldering versus auto assembly

- Custom parts from a supplier who provides a prototype using a different manufacturing process or tooling other than what will be used in the final production version
- Socketed parts versus nonsocketed parts
- Printed Circuit Board (PCB) fabricated from a small quick-turn facility versus the standard fabricator.

There can be another problem with using different suppliers for the prototype. Suppose Supplier A provides a quick-turn delivery and is used for prototypes. Supplier B is used for production. If Supplier A finds a problem in the design and fixes it on his print, communicates it to the designer, but the designer fails to update his documents, then the problem may surface again in production, and there will be a scramble to figure out why. In PCB fabrication, manufacturers use different (often custom) software programs to check for layout problems in the board artwork. These problems affect the yield in PCB manufacturing, and the manufacturer will fix them. The PCB manufacturer considers this part of the fabrication service and routinely "cleans up" your design. However, the supplier may not notify you of the fixes. After the material is tested and found to be acceptable, it is transferred to a PCB manufacturer who will be used for volume production. The design issues were not fixed in the artwork, so problems surface in production that were not identified with the prototype. This kind of problem can be avoided. Do not allow suppliers to make any changes to the artwork or the design without submitting a timely engineering change request.

The assemblies for HALT testing should be built using the standard manufacturing process that will be used in production. Any special tooling required for production should be utilized in the assembly of the boards for test. Do not build the boards for HALT by hand in a prototype lab. Build test assemblies the same way you would production assemblies. HALT will reveal manufacturing as well as design-related problems. Therefore, it is best to mimic as close as possible the design and manufacturing process.

Before you HALT the product, some final planning is in order. There is a significant amount of preparation work required before HALT. The reliability engineer and the design engineer must ensure that everything is in place before HALT begins. The following is a list of items that have to be ready before the test:

1. The product that will be tested, five working units, and one more which is considered golden. The "golden unit" is not stress tested; it is used when there are testing issues or problems. The golden unit is used to identify if the problem is in the product under test or the test instrumentation. Inserting the golden unit will verify if the problem exists with the product. This will speed the troubleshooting process greatly. The golden unit is also used in troubleshooting, because the failed product can be compared to

the golden unit to narrow the source of the failure. Hence the information learned by using this unit is truly "golden."

2. Test instrumentation. This is probably the most important item on the list after the product itself because failures have to be discovered and corrected. Poor monitoring will miss some failures and render the HALT process less effective than it might otherwise have been.

3. The output specifications that will be monitored and the monitoring instrumentation

4. Documentation, such as, schematics, assembly drawings, flowcharts, and so on

5. A mechanical fixture to affix the product to the HALT table

6. Input and output cabling

7. Input and output liquid cooling hoses

8. Special devices, that is, liquid cooling apparatus, chillers, air ducts, power sources, other support devices, and so on

9. Test software if required

10. The stress levels intended to be applied to the product (established and agreed to by the HALT team)

11. Scheduling the required time for testing

12. The responsible design engineer is available to support the test for the entire HALT process

13. A test engineer is available to debug test instrumentation problems and to assist in failure analysis for the entire HALT process

14. The reliability engineer and a HALT chamber operator are available to support the test for the entire HALT process. (The reliability engineer documents the tests and writes the final HALT test report.)

After all the preparation work is complete, it is time to start HALT to precipitate failures. The tests can be as short as a few days or as long as several weeks. The length of the test is dependent on how many assemblies are available to test, the time it takes to fix failures, the frequency at which failures occur, and the frequency and length of downtime due to test setup and equipment problems.

It is recommended that you have six assemblies for the HALT test. One assembly is a "golden unit" and does not get stress tested. You can do HALT testing with less assemblies but the process will take longer and you may not identify fewer design issues. Use the "golden unit" to check out the system and as a troubleshooting aid when failures occur. The other five units are used for HALT testing. Place the first assembly into the chamber and orient it in a way to efficiently transfer the chamber air to the assembly. Then secure it to the chamber's vibration table. Use locking hardware to secure the assembly to the chamber. If locking hardware is not used, there is a risk that the assembly will loosen under vibration. Torque to specification all hardware with a calibrated torque driver. Connect power and I/Os to the system and secure them to the chamber (the cables should move with the vibrating table). After everything is set up, run a baseline ambient test to verify proper operation (a minimum of ten minutes or as long as it will take to run the diagnostics two times). Next run a tickle vibration test (5 Grms) to verify that there are no loose electrical or mechanical connections. If everything passes, then you are ready to start the HALT testing.

There may be protection circuitry in the assembly that prevents the product from operating above a threshold point – for example temperatures above 85 °C. This type of circuitry may need to be disabled for HALT (unless it is for a safety issue). DC converters typically have thermal shutdown circuitry to prevent them from operating at temperatures above their specified maximum value. If the protection circuitry is needed or embedded in a component, then a strategy can be developed to locally control the temperature of that device so the rest of the product can be tested above the protection point. (You should wait to verify that the protection circuitry works before disabling it in the system.)

16.2.1 Starting the HALT Test

Once everything is in place for HALT, it is time to begin the test. HALT testing requires the device to be operational during the test. There is no reason to perform HALT if the device cannot be tested under operational conditions. Passive stress testing reveals little to no useful information about the design and is ineffective at precipitating failures. The HALT process flow is illustrated in Figure 16.2.

The HALT test begins with single stresses and is followed by combinational stress tests. We have suggested an order in which the testing proceeds, but there is nothing wrong with changing the order. It is recommended that testing begin with single stresses first before proceeding to combinational stress testing. A typical testing sequence for HALT is as follows:

1. Room ambient
2. Tickle vibration test
3. Temperature step stress test
4. Rapid thermal cycling stress test
5. Vibration step stress test
6. Combinational temperature and vibration test
7. Combinational search pattern test
8. Additional stresses (these are optional and based on the product and use environment) are line voltage and frequency margining, power sequencing, clock frequency, load variation, and so on.

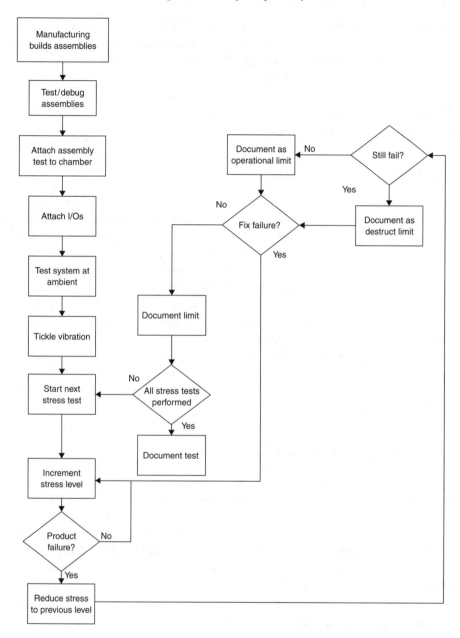

Figure 16.2   HALT process flow

The HALT test starts with placing the product into the HALT chamber and securing it mechanically to the chamber. The mechanical structure should be stiff, strong, and lightweight. The purpose of the fixturing is to firmly secure the product under stress without adversely affecting the test. Test fixtures that are heavy in mass complicate a rapid thermal cycling test and require longer dwell times for a product to stabilize. Next, connect the I/O (inputs and outputs) to the product (i.e., power sources, loads, and instrumentation). Ensure that the I/Os are firmly attached to the product. Finally, attach the stress monitoring sensors to the product (i.e., accelerometers and thermocouples). The accelerometers can be attached using Super Glue® or any other type of Cyanoacrylate adhesive. The thermocouples are attached using either a thermally conductive adhesive or Kapton® adhesive tape.

16.2.2   Room Ambient Test

Once everything is secured to the chamber, the doors are closed and the test chamber is turned on. Set the chamber to room ambient and perform a nitrogen purge to evacuate any moisture residing in the chamber. The product is turned on and allowed to stabilize before performing the first diagnostic test to verify everything is operating properly. This step verifies that the test setup is functioning properly. The test time varies based on how long it takes to verify that everything is functioning properly. It is often a function of the time it takes the test software to run a complete functional test. The test software should achieve 100% test coverage. Being able to run complete functional testing is desirable but not always possible. When complete test coverage is not possible, there need to be assurances that there is at least sufficient test coverage to determine if the product is operating properly or a second HALT is performed when test software is complete.

16.2.3   Tickle Vibration Test

After the test setup, HALT chamber and product are confirmed to be operating properly, a test is run to ensure that all mechanical connections are secure. This is achieved using a tickle vibration test. The tickle vibration test applies low-level vibration between 2 to 5 Grms. The chamber temperature is set to maintain room ambient temperature and functional testing of the product is repeated to verify that the device is mechanically secure and that there are no loose connections.

16.2.4   Temperature Step Stress Test

Before starting the temperature step stress test, the upper and lower stress physical limits in temperature are defined. These limits represent the point at which the material changes physically (often referred to as a phase change) and results in a failure. An example of this would be a connector with a plastic body that melts or becomes soft when a temperature threshold is reached. Failure at this temperature is expected and represents limitations in the design and not a product failure.

The temperature step stress starts at ambient and proceeds to lower temperatures in typically 10 °C increments. Testing starts with the weakest stress and moves to stronger stresses as the testing continues. This way the subtle failures will not be lost with excessive stress testing. Once the temperature is reached, the product dwells for typically 10 to 15 min. The dwell time includes the time required to run functional tests to verify proper operation. Continue in steps until the lower "cold" temperature limit is reached. After reaching the lower temperature limit, the product is returned to room ambient temperature and functional testing is performed to verify that the product is operating correctly. The product now begins temperature step stress until the upper "high" temperature limit is reached.

The temperature step stress test continues until a failure is precipitated. Record the point where the failure occurs. Now reduce the stress to the previous stress level, to find out if the system recovers. If the system begins to work again, then the failure is identified as a soft failure. If the system does not recover, then the failure is identified as a hard failure. Document the failure and the stress level that caused it.

There may be the possibility to "band-aid" the failing element in order to continue increasing the stress. If the fix is simple, it may be possible to implement while the product is in the chamber. If troubleshooting is required, remove the product and place the next product in the chamber for testing. One advantage of having five units for test is that HALT testing can continue while the recently failed unit is being fixed.

After the high temperature has been reached, the product is returned to room temperature. At the end of this test, the upper and lower soft temperature limits (soft failure) and the upper and lower destruct limits (hard failure) have been identified. The first stress test, temperature step stress, is now complete. Figure 16.3 shows this test graphically.

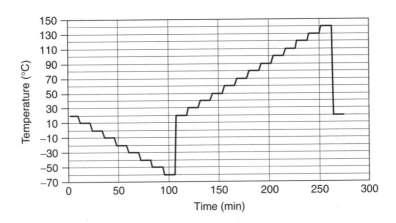

Figure 16.3   Temperature step stress

16.2.5   Rapid Thermal Cycling Stress Test

The next stress test is rapid thermal cycling. The upper and lower operational limits (soft failure) were identified in the previous test. In the rapid thermal transition test, the product is rapidly transitioned to just below the upper and lower operational limits. In general, keeping the temperature limits to 5 °C below the operational limits is sufficient. The chamber temperature is made to change as rapidly as possible. Once the product reaches ramp temperature, it is allowed to dwell there typically for 10 to 15 min so that the product reaches that temperature before ramping to the next set point. If the dwell times are not long enough for the product to stabilize at the temperature, the product will see a lot less stress during temperature ramp. This test method uncovers the extreme thermal rate of change weaknesses. Running several rapid temperature excursions, between three to five cycles is sufficient. The rapid thermal cycling test is shown graphically in Figure 16.4.

16.2.6   Vibration Step Stress Test

The next stress test is a vibration step stress that is applied to the product. In this test, the product is maintained at ambient temperature while vibrational stresses are increased in 5 to 10 Grms increments. The test continues until the limit of the chamber's capability is reached, the upper vibrational limit is reached, or the product can no longer survive higher stress levels. With each step in vibration, the product is tested to verify proper operation. When stress levels exceed 20 Grms, it may be necessary to run a tickle vibration to detect failure. Many times vibration-caused failures do not reveal themselves to the test instrumentation at the higher vibration levels, but the failure becomes apparent at the lower levels. Document failures and troubleshoot to root cause. The vibrational step stress test is shown graphically in Figure 16.5.

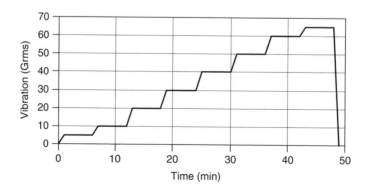

Figure 16.4   Rapid thermal cycling (60 °C/min)

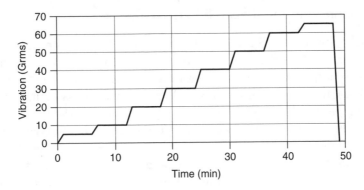

Figure 16.5   Vibration step stress

16.2.7   Combinational Temperature and Vibration Test

After testing has been completed for single types of stresses, combinational stresses are applied. The first combinational stresses are temperature and vibration. In this test, the temperature starts at ambient and is stepped in 10 to 20 °C increments until just below the upper and lower destruct limits are reached. The product remains at each temperature stress, while vibrational stresses are induced on the product in 10 to 20 Grms increments. At each vibration stress level, the product is allowed to stabilize and functional testing is performed.

It is very important to record the stress levels at which the soft and hard failures occurred. Later, when you have made design corrections these stress levels should have increased, thus increasing the products operating margins. The combinational temperature and vibration test is shown graphically in Figure 16.6.

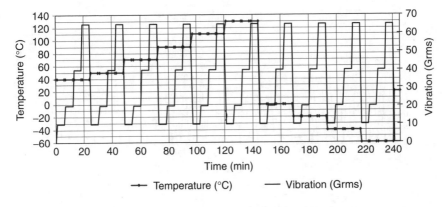

Figure 16.6   Temperature and vibration step stress

16.2.8   Combinational Search Pattern Test

A relatively new HALT technique created by Dr. Greg Hobbs is the "search pattern technique." The idea is to slowly sweep temperature and rapidly sweep vibration simultaneously. Starting with the product at room temperature (or about 25 °C), the temperature is lowered to the lower stress limit, say −40 °C. At the same time, vibration is sweeping as fast as it can between 0 to 20 Grms. Typically, the vibration will go from the low level to the high level and back down again in less than 30 s (this is adjustable in some HALT chambers). Once the vibration stresses are started, the temperature is slowly swept from −40 °C to +140 °C (hypothetical values) and then back to room temperature. If the temperature rate of change is set to 2 °C per minute, the entire test will take 4.4 h (refer to Figure 16.7).

The search pattern technique is valuable where the soft failure is very close to the hard failure. The temperature changes slowly while the product is being continuously monitored. This allows the test to be stopped before a hard failure is encountered. This opens opportunities for some failure investigation before the hard failure is found. Another advantage of slowly sweeping temperature overtemperature step stress is that it will reveal any oscillations or instabilities that occur only at a specific temperature point or narrow range. If you use step stresses, there is a possibility that you will pass over the point of instability.

16.2.9   Additional Stress Tests

Depending on the product, there may be additional stresses that are appropriate. Some additional stresses that can be applied are DC power supply voltage margining (first single supplies then different supply voltages in combination),

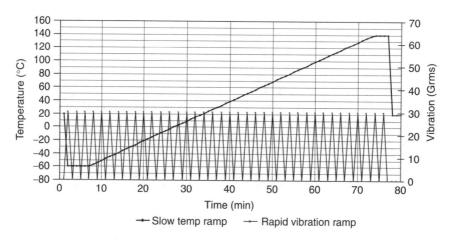

Figure 16.7   Slow temperature ramp and sinusoidal amplitude vibration

AC line input voltage and frequency margining, timing margining, output loading, clock oscillator frequency variation, power cycling, and power sequencing.

16.2.10 HALT Validation Test

During HALT, failures will surface. Each HALT failure is documented either through a FRACAS system or in some other form to log failures (see Figure 16.8). Some of the failures will be fixed while the unit is in the chamber. Others may require a "band-aid" fix so that testing can continue. Often, components that have failed are removed which will require failure analysis to determine root cause. Getting to root cause for all failures is one of the requirements of HALT. After HALT is completed, there will be a list of failures identified along with the root cause and the stress required to precipitate the failure. Ideally, everything that fails is fixed through design changes. However, this is not always practical. Each design change has an associated economic and schedule impact that can be weighed against the improvement in design margin, reliability, and first-pass yield. The level of stress required to precipitate the change can also play a role in the decision to fix a particular failure. There should be a commitment to fix all failures through design change except when it can be shown to not make economic or business sense.

After all the agreed upon design changes have been implemented, a final validation HALT test is performed to verify the effectiveness of the design

HALT FORM

Assembly # _____ Rev: ____ S/N: _____   Page___ of___

Date ___/___/___   Owner: _____

HALT Team Members _____

			ACTION ITEM LIST & CORRECTIVE ACTION RECOMMENDATIONS									
Item #	PROBLEM or ISSUE	CAUSE	EFFECT	T	G	V	Sf	RPN	RECOMMENDED CORRECTIVE ACTION	WHO	WHEN	A

HALT STRESS LEVELS						
STRESS LEVELS - RECOMMENDED Other (Enter Value)	DESIGN LIMITS ( )	10% OVER DESIGN ( )		20% OVER DESIGN ( )		>30% OVER DESIGN ( )
WEIGHT FACTOR (T,G,V)	10	5		3		1
VIBRATION STRESS:						
TEMPERATURE STRESS:						
VOLTAGE STRESS						

Note any deviations from HALT Guideline: _____

Legend:
T = Temperature in °C;   G = Vibration in Grms;   V = Voltage margin level;   Sf = Safety hazards;   RPN = Risk priority number;   A = Audit

Figure 16.8 HALT form to log failures

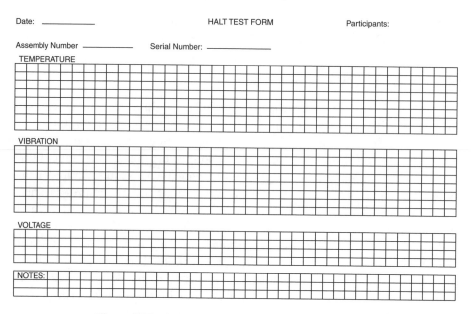

Figure 16.9 HALT graph paper for documenting test

changes and to ensure no new failures were injected into the product. This test can be on a single device, although testing several is desirable. In addition, the validation HALT test doesn't need to be as rigorous as the original HALT; increasing the stress level increments will shorten the test time.

Finally, keeping track of all the testing that was performed in HALT can be difficult. There are many different types of tests performed and the sequence can vary. In addition, keeping track of which units were tested when, test equipment failures, and test anomalies can be challenging. Developing a form to track all this activity will be a lifesaver later when it is time to write the HALT report. A sample form can be found in Figure 16.9.

16.3 PROOF OF SCREEN (POS)

During the HALT test, the product's soft and hard failure limits were identified. The HALT limits determine the appropriate Highly Accelerated Stress Screen (HASS) that will be used to weed out manufacturing defects in production. HASS is described in detail in Chapter 7. HASS applies accelerated stress levels to the product so that process-related defects are precipitated to fail. HASS replaces traditional burn-in or other forms of Environmental Stress Screens (ESS) because it is more efficient at removing process defects and is less damaging to the product life (cumulative stress from burn-in is less).

The HASS profile consists of two parts, the precipitation screen and the detection screen. The test begins with the precipitation screen. The precipitation

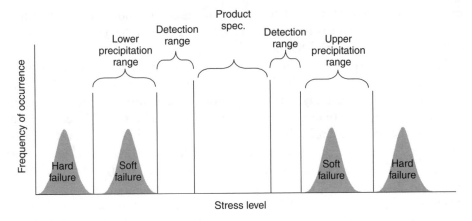

Figure 16.10   HASS stress levels

screen is a stress level that is below the destruct limit and above the operating limit. The precipitation screen accelerates process defects to failure (refer to Figure 16.10). The HASS screening level applied to the product needs to be determined. A good stress level for temperature is between 80% and 50% of the destruct limits. The initial vibration stress level is at 50% of destruct limits. It is important to stay below the destruct limits; otherwise damage to a good product is likely. The purpose of the precipitation screen is to sufficiently damage defective products so it can be detected in test. Bad products are identified as the assembly passes through the detection screen.

During detection screening, the temperature stress is reduced to levels below the soft failure limit but above the product specification limit. The HASS profile is usually short, typically three to five cycles of precipitation and detection is adequate to detect failures before they occur during customer use.

Detection and precipitation screen is performed as one operation. You increase temperature past the soft failure range but below the damage level (this is the precipitation phase), then the stress is reduced below the soft failure level where a "good" assembly will recover (this is the detection range). If it doesn't recover, then a defect has been detected.

The HASS profile must not damage or severely degrade good products. Generally, if the right stress levels are applied, the defective assemblies will degrade at a significantly greater rate than good products so that they can be easily detected. The HASS profile must also be severe enough to precipitate a process defect to failure. A Proof Of Screen (POS) is used to ensure that the HASS levels are not damaging good product by removing too much product life but effective enough to identify defective units.

The environmental stresses induced on the product by the HASS screen will remove some of the life expectancy of the product. This is unavoidable. The goal of the HASS screen is to provide a stress level high enough to precipitate manufacturing defects without removing an excessive amount of product life. How much product life is removed in HASS can be estimated through a process called "Proof Of Screen" (POS).

The POS process repeats the HASS screen over and over again on a good product until it fails. Each HASS screen removes some product life. Applying the HASS stress repeatedly causes the product to continually degrade at an accelerated rate. Eventually, the product will fail because the stress continually degrades the product until it has reached wear out. If it takes 20 HASS cycles to render the product nonoperational, then it is reasonable to estimate that 5% of the product life is removed with each HASS screen. If the device failed after only four HASS screens, it can be assumed that 25% of the life of the product was removed each time. There is no minimum number of stress cycles desired before a product fails. Some companies want at least 20 cycles without a failure. The test should be run on a large enough sample size to ensure that normal manufacturing process variations are accounted for.

If, on the other hand, you run the HASS test for 100 cycles without a failure, the HASS stress levels may be set too low. Some practitioners recommend seeding products when doing POS to determine if the HASS screen is effective at detecting defective products. Seeding a board requires intentionally inserting a known manufacturing defect(s) into the product. The product undergoes HASS to determine if the defects can be found during the HASS screen. The problem with seeded defects is that it is difficult to insert seeded defects that are real representations of product defects (i.e., manufacturing process drift or supplier changes).

16.4 OPERATE FRACAS

A Failure Reporting, Analysis and Corrective Action System (FRACAS) was installed in the previous phase. The FRACAS database (often a purchased software program) is customized for the user's particular application. The customization of the database is part of the installation and checkout process for FRACAS.

With the product now in the design validation phase, prototypes are built and tested to validate the performance of the product. During development testing and design validation, failures will occur in the product. There is a natural tendency, especially early in the development phase, to treat these failures informally. They are often noted in a notebook, a piece of paper, or sometimes are fixed without any documenting at all. Treating any failure as irrelevant is shortsighted no matter how insignificant the failure may seem. These failures often resurface later in the program when it is more costly to rectify.

FRACAS prevents this from occurring. FRACAS becomes operational once the first prototypes are built. Every failure that occurs from that time on is recorded in the FRACAS database. The failures are recorded as well as the activity to determine the root cause and corrective action. FRACAS will track the progress being made to resolve failures as well as indicate the rate at which new failures are occurring. This information will provide a good indication as to how fast the design is maturing or if it is still in a state of flux.

For FRACAS to be effective, everyone who identifies a failure must use it. Preventing designers from using their own system to document failures during prototype can be a difficult challenge. The problem can be resolved by providing training in the previous phase on how to use the FRACAS database. This, in conjunction with a commitment from the management that all designers will use the FRACAS database to record every failure, will ensure success.

16.5   DESIGN FMEA

In the product design phase, an FMEA was performed on every subassembly, circuit board, and at the system level. There was a significant amount of effort and resources required to complete the task. The FMEA in the design validation phase is not intended to repeat the previous effort but to complement it. The design validation FMEA complements the previous effort by only evaluating significant design changes (ECOs) that resulted out of design validation, HALT, and other design-related failures. AN FMEA is not required for simple ECOs, such as a component value change. Significant ECO changes usually result in a new board or mechanical layout. The design FMEAs that are conducted in the design validation phase only address that which has changed. The FMEA is not repeated for the entire assembly; only significant changes made to the design need to be analyzed using with an FMEA. The time required to perform an FMEA for a design change will be significantly less than the time required for the original FMEA.

16.6   CLOSURE OF RISK ISSUES

At the end of the design validation phase, the product is complete and ready for market. The high-risk issues captured earlier should be closed before the end of the design validation phase. There should be no unresolved high-risk issues; it doesn't matter if the risk is a design, manufacturing, test, supplier, or a reliability issue. Any unresolved high-risk issues represent escapes in the risk mitigation process. Each high-risk issue has a contingency (backup) mitigation plan that should resolve the risk issue before entering the production phase. If a high-risk issue has not been resolved before entering the production phase, escalation is required.

Escalation of unresolved high-risk issues is required because these issues often become costly problems once a product is on the market. The escalation process starts well before the completion of the design validation phase. Escalation begins by elevating the problem to senior management. Often, these problems are related to the way in which the risk is being managed, the type of resources used to solve the problem, or the skills of the people working to fix the problem. Senior management must determine why the problem is not being resolved and implement changes to fix it.

At the end of the design validation phase, the product is ready to be sold.

References

FMEA

1. *Recommended Failure Modes and Affects Analysis (FMEA) Practices for Non-Automobile Applications*, SAE (2001).
2. M. Krasich, *Use of Fault Tree Analysis for Evaluation of System Reliability Improvements in Design Phase*, 2000 Proceedings Annual Reliability and Maintainability Symposium (2000).
3. K. Onodera, *Effective Techniques of FMEA at Each Life-Cycle Stage*, 1997 Proceedings Annual Reliability and Maintainability Symposium, IEEE (2000).
4. S. Bednarz, D. Marriot, *Efficient Analysis for FMEA*, 1998 Proceedings Annual Reliability and Maintainability Symposium (1998).
5. M. Kennedy, *Failure Modes and Effects Analysis (FMEA) of Flip-Chip Devices Attached to Printed Wiring Boards (PWB)*, IEEE/CPMT International Manufacturing Technology Symposium, IEEE (1998).
6. R. Whitcomb, M. Riox, *Failure Modes and Effects Analysis (FMEA) System Development in a Semiconductor Manufacturing Environment*, IEEE/SEMI Advanced Semiconductor Manufacturing Conference, IEEE (1994).
7. D. J. Russomanno, R. D. Bonnell, J. B. Bowles, *Functional Reasoning in a Failure Modes and Effects Analysis (FMEA) Expert System*, 1993 Proceedings Annual Reliability and Maintainability Symposium, IEEE (1993).
8. S. Prasad, Improving Manufacturing Reliability in IC Package Assembly Using the FMEA Technique, *IEEE Transactions of Components, Hybrids and Manufacturing Technology*, 14(3), 452–456 (1991).

Acceleration methods

1. H. Caruso, A. Dasgupta, *A Fundamental Overview of Accelerated-Testing Analytic Models*, 1998 Proceedings Annual Reliability and Maintainability Symposium, IEEE (1998).
2. J. Evans, M. J. Cushing, P. Lall, R. Bauernschub, *A Physics-of-Failure (POF) Approach to Addressing Device Reliability in Accelerated Testing of MCMS*, IEEE (1994).
3. P. Lall, Tutorial: Temperature as an Input to Microelectronics-Reliability Models, *IEEE Transactions on Reliability*, 45(1), 3–9 (1996).
4. M. J. Cushing, *Another Perspective on the Temperature Dependence of Microelectronic-Device Reliability*, 1993 Proceedings Annual Reliability and Maintainability Symposium (1993).

ESS

1. S. M. Nassar, R. Barnett, *Applications and Results of Reliability and Quality Programs*, 2000 Proceedings Annual Reliability and Maintainability Symposium, IEEE (2000).
2. H. Caruso, A. Dasgupta, *A Fundamental Overview of Accelerated-Testing Analytic Models*, 1998 Proceedings Annual Reliability and Maintainability Symposium, pp. 389–393 IEEE (1998).
3. G. A. Epstein, *Tailoring ESS Strategies for Effectiveness and Efficiency*, 1998 Proceedings Annual Reliability and Maintainability Symposium, 37–42, IEEE (1998).
4. H. Caruso, *An Overview of Environmental Reliability Testing*, 1996 Proceedings Annual Reliability and Maintainability Symposium, IEEE (1996).
5. M. R. Cooper, *Statistical Methods for Stress Screen Development*, 1996 Electronic Components and Technology Conference, IEEE (1996).

HALT

1. General Motors Worldwide Engineering Standards, *Highly Accelerated Life Testing*, GM (2002).
2. G. K. Hobbs, *Accelerated Reliability Engineering*, John Wiley & Sons (2000).
3. H. W. McLean, *HALT, HASS & HASA Explained: Accelerated Reliability Techniques*, American Society for Quality (May, 2000).
4. J. Strock, Product Testing in the Fast Lane, *Evaluation Engineering* (March, 2000).
5. N. Doertenbach, *High Accelerated Life Testing – Testing With a Different Purpose*, IEST, 2000 proceedings (February, 2000).
6. D. Rahe, *The HASS Development Process*, 2000 Proceedings Annual Reliability and Maintainability Symposium, IEEE (2000).
7. D. Rahe, *The HASS Development Process*, ITC International Test Conference, IEEE (1999).
8. M. Silverman, *Summary of HALT and HASS Results at an Accelerated Reliability Test Center*, Qualmark Corporation, Santa Clara, CA, 1998 Proceedings Annual Reliability and Maintainability Symposium, IEEE (1998).
9. M. Silverman, *HASS Development Method: Screen Development, Change Schedule, and Re-Prove Schedule*, 1998 Proceedings Annual Reliability and Maintainability Symposium, IEEE (1998).
10. R. H. Gusciaoa, *The Use of Halt to Improve Computer Reliability for Point of Sale Equipment*, 1998 Proceedings Annual Reliability and Maintainability Symposium, IEEE (1998).
11. J. A Anderson, M. N. Polkinghome, *Application of HALT and HASS Techniques in an Advanced Factory Environment*, 5th International Conference on Factory 2000 (April, 1997).
12. M. L. Morelli, *Effectiveness of HALT and HASS*, Hobbs Engineering Symposium, Otis Elevator Company (1996).
13. C. Ascarrunz, *HALT: Bridging the Gap Between Theory and Practice*, International test Conference 1994, IEEE (1994).

14. R. Confer, J. Canner, T. Trostle, S. Kurz, *Use of Highly Accelerated Life Test Halt to Determine Reliability of Multilayer Ceramic Capacitors*, IEEE (1991).
15. H. McLean, *Highly Accelerated Stressing of Products with Very Low Failure Rates*, Hewlett Packard Co., (1991).
16. P. E. Joseph Capitano, Explaining Accelerated Aging, *Evaluation Engineering*, p. 46 (May, 1998).
17. E. R. Hnatek, Let HALT Improve Your Product, *Evaluation Engineering*.
18. G. K. Hobbs, What HALT and HASS Can Do for Your Products, *Evaluation Engineering*.
19. G. K. Hobbs, What HALT and HASS Can Do for Your Products, *Hobbs Engineering, Evaluation Engineering*, Qualmark Corporation, p. 138 (November, 1997).
20. E. O. Minor, *Quality Maturity Earlier for the Boeing 777 Avionics*, The Boeing Company.
21. M. A. Silverman, *HALT and HASS on the Voicememo II™*, Qualmark Corporation.
22. M. Silverman, *Summary of HALT and HASS Results at an Accelerated Reliability Test Center*, Qualmark Corporation, Santa Clara, CA.
23. M. Silverman, *Why HALT Cannot Produce a Meaningful MTBF Number and Why this should not be a Concern*, Qualmark Corporation, ARTC Division, Santa Clara, CA.
24. W. Tustin, K. Gray, Don't Let the Cost of HALT Stop You, *Evaluation Engineering*, pp. 36–44.

17

Production Phase

There are two major objectives in the production phase. The first addresses production ramp (Table 17.1). The objective here is to quickly achieve design maturity and ramp to the desired production levels. The second major objective is to have mechanisms in place (quality controls) to ensure the quality of the product before volume production is achieved. The activities that take place in the production phase are illustrated in Table 17.2.

17.1 ACCELERATING DESIGN MATURITY

When a product goes into production, there will inevitably be problems. These problems impede the ability to achieve volume production including the ability to quickly increase or decrease volume production. The problems affect manufacturing and test yields and are composed of both design and process issues. These problems are the escapes in the product development process that occurred due to inadequate execution of the design and reliability process. It is often a result of either not doing or failing to close in on issues identified in risk mitigation, Highly Accelerated Life Test (HALT), Device Verification Test (DVT), Failure Modes and Effects Analysis (FMEA), and Highly Accelerated Stress Screens (HASS). These reliability tools, when implemented correctly, enable a product to quickly achieve design maturity.

The design is considered mature when the inherent reliability of the design is achieved. Typically, early in production, design, manufacturing, test, and reliability problems surface, which are significant enough to require a design change to fix them. After all these fixes are implemented, the product begins to achieve the reliability goal. At this point, the design has reached maturity and the quality/reliability issues that surface are few and far between. The problem most companies have in reaching design maturity is that they take

Improving Product Reliability: Strategies and Implementation.   Mark A. Levin and Ted T. Kalal
© 2003 John Wiley & Sons, Ltd   ISBN: 0-470-85449-9

Table 17.1   Reliability Activities in the Production Ramp Phase

Participants	Product ramp phase		
	Reliability activities	Deliverables
• Reliability engineering	1. All products have HASS until acceptable pass rate is achieved.	1. HASS pass-fail report.
	2. Operate FRACAS. All product and customer failures entered into FRACAS database and tracked to closure.	2. Periodic FRACAS and FRB meeting. Failures tracked to closure.
• Design engineering	3. Start reliability growth. Product operating time and failure events entered into reliability growth chart. Progress at removing failure modes is evaluated.	3. Reliability growth curves.
• Manufacturing engineer	4. FMEA is performed (on any significant design changes only) and process FMEA.	4. Completed FMEA spreadsheet and closure on corrective action items.
• Test engineering	5. SPC Program initiated. Production yield is monitored and adjusted as needed.	5. Production quality data reported.
• Field service/customer support		
• Purchasing/supply management		
• Safety & regulation		

Note: FMEA: Failure Modes and Effects Analysis; FRACAS: Failure Reporting, Analysis, and Corrective Action System; HASS: Highly Accelerated Stress Screens; SPC: Statistical Process Control; FRB: Failure Review Board.

too long to identify these issues and even longer to identify root cause and corrective action.

By the time a product reaches volume production, the design must be in a mature state. When a design reaches maturity, the majority of engineering resources are no longer needed to support the product. These vital engineering resources can then be directed toward their primary goal, developing new products to increase market share, and expand the business. Achieving design maturity takes time, but it is accelerated by the reliability process.

All too often, it takes between two to five years for a new product to reach design maturity. Initially, the product starts out with a high failure rate [low Mean Time Between Failures (MTBF)]. When the product reaches maturity, the failure rate flattens out to what the design MTBF is capable of achieving.

Table 17.2   Reliability Activities in the Production Phase

Participants	Product phase	
	Reliability activities	Deliverables
• Reliability engineering	1. Switch from HASS to HASA once production pass rate is achieved.	1. HASA pass-fail report.
	2. Operate FRACAS. All product and customer failures entered into FRACAS database and tracked to closure.	2. Periodic FRACAS and FRB meeting. Failures tracked to closure.
• Manufacturing engineer	3. Continue reliability growth. Product operating time and failure events entered into reliability growth chart. Progress at removing failure modes is evaluated.	3. Reliability growth curves.
• Test engineering	4. FMEA is performed (on any ECOs only) and process FMEA.	4. Completed FMEA spreadsheet and closure on corrective action items.
• Field service/customer support	5. SPC Program continued. Production yield is monitored and adjusted as needed.	5. Production quality data reported.
• Purchasing/supply management		
• Safety & regulation		

Note: HASA: Highly Accelerated Stress Audit; ECO: Engineering Change Order.

At this point, design-related problems rarely surface; the product reliability stops improving in value (MTBF). At this point, the product has reached its achievable MTBF. Assuming that a lot of variability is not there in the manufacturing process, the first-pass yield in production will flatten out as well. Simply stated, when a product reaches design maturity, there is little more that can be done to further improve the quality and reliability of the product, without significantly changing the design.

The time it takes for a product to reach design maturity is important because this is when the product has reached its lowest cost structure. Therefore, it should be easy to understand why there is such a big push early in the production phase to reach the product rampability and yield targets. Product life cycles will continue to shorten and technology will drive product obsolescence at a greater rate. If the time it takes to reach design maturity doesn't also reduce, the products developed can become obsolete before they ever reach design

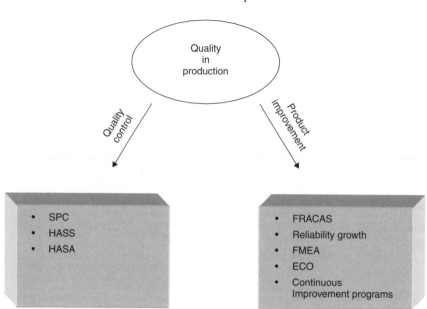

Figure 17.1   Achieving quality in the production phase

maturity. Fortunately, the reliability process we have presented is the most effective way to accelerate the maturity of a design.

All the hard work in the product development phase has resulted in a reliable and robust design. The product has been designed to be reliable, but this alone is no guarantee for success. There is still more work to be done. Poor quality control in manufacturing and test will make the product unreliable. However, achieving high quality in manufacturing is not difficult. Only a few tools are needed to ensure quality in the products produced. These tools fall into two categories, quality control and product improvement as illustrated in Figure 17.1.

17.1.1   Product Improvement Tools

As mentioned earlier, the first and foremost objectives during production ramp are to achieve volume production and design maturity as quickly as possible. Many problems can surface after product release that vary in severity. Some affect the customer, that is, "Out-of-Box" failures (also referred to as *dead on arrival*), installation problems, and high failure rate. Other problems affect manufacturing, that is, low first-pass yield, significant rework and scrap, and high boneyard pile (faulty product waiting to be fixed).

The product improvement tools are designed to solve these problems by focusing on data collection and data trending to detect early problems in the product. Each issue is then driven to root cause and corrective action. When issues arise that have significant impact but the problem resolution is not easy, a containment plan is put in place as a short-term fix. The containment solution is often costlier than the design fix. These same tools are then used to verify that the fix was effective.

FRACAS

Failure Reporting, Analysis, and Corrective Action System (FRACAS) began operating in the design validation phase. As product failures surfaced from prototype testing, design validation, and HALT, they were entered into the FRACAS database and their progress was tracked to root cause and corrective action. By the time the design is transferred to production, the problems that surfaced during design validation should be resolved, validated, and implemented. The FRACAS database provides the ability to make that determination.

Once the design enters production, the FRACAS system changes. During design validation, with the exception of customer alpha and beta testing, the product development team identified all the failures. Once the product is in production, the failure reporting data can come from many different places. Internally, failure data can come from assembly, test, receiving inspection, component engineering, supplier management, and reliability (the list goes on). In addition, failure reporting can also come from outside sources such as the customer, customer service and support, field service and repair, and marketing. The fact that failure reporting information comes from so many different sources can be a problem.

When failure reporting is coming from many different sources, there is a good likelihood that the data will be entered and stored in different places. FRACAS is an extremely effective tool in identifying failure patterns, so problems can be identified at the earliest possible point. If the data is stored in different databases, then it becomes difficult to impossible to gather the data needed to identify the trend. This is often one of the biggest problems with FRACAS. Different groups have different systems that are used for failure reporting. Getting everyone to use the same FRACAS system can be difficult to accomplish; often, groups are unwilling to change systems either because of the cost, time, and inability to perform their needs or the inability to transfer the existing database to the new format. All these issues are valid problems; the sooner they are resolved, the better.

Another problem common to FRACAS systems is in the consistency of the data. As we have pointed out, failure reporting can come from many different sources. The quality of the reporting data entered into FRACAS is often a source of problems. Some of the common problems with recording failures into the FRACAS database are incomplete data, insufficient failure information, inconsistent failure reporting (the same failure is described differently so the magnitude of the failure may not be known), and failure data is not entered. These issues can be minimized and some can be eliminated entirely through a well-structured FRACAS system.

Data is entered into the FRACAS database on a continuous basis. This data is analyzed for failure trends and severity. However, just because there is a process to identify and track failures, it doesn't ensure that these failures will be resolved, let alone in a timely fashion. In order for a FRACAS system to be effective, it requires a Failure Review Board (FRB) to oversee problem resolution. The FRB board consists of a team of individuals who are responsible for the product and have the authority to resolve it. The FRB board consists of the following members:

- FRB leader (senior level manager)
- Design team representative (others may be pulled in)
- Reliability engineer
- Manufacturing representative.

This group represents the minimum team participation. The FRB team meets on a regular scheduled basis, often weekly. Attendance at the meeting is mandatory, so, if a team member cannot make it, their assigned backup fills in. The task of the FRB team is to review the failure reports, problem severity [FRACAS systems can assign design severity similar to the Risk Priority Number (RPN) process used in FMEAs], and to prioritize problems. The FRB board then assigns appropriate team members who will be responsible for the top issues. These individuals resolve top issues to a root cause, determine corrective action, and implement fix. The FRACAS database is then used to validate if the fix was effective. An FRB member, who is assigned responsibility to resolve a problem, must have the authority to implement the needed changes. If manufacturing is assigned to fix a problem that requires an engineering design change, it is unlikely that they will be successful. Design issues should be assigned to the design team to fix; it is not for manufacturing to find a way to "band-aid" the problem. (As the top issues are resolved, the lower issues become the new top issues and so on.)

Design issue tracking

A very simple alternative approach to FRACAS that tracks nonconformances to closure is design issue tracking. This is nothing more than an "Action Item" list with a stacked bar chart to graph the progress of the activity. As can be seen from Figure 17.2, the graph tracks failure bugs from documentation, to being assigned, to failure resolution and validation, and finally to closure. The height of the bars (y-axis) will stop increasing when new problems no longer surface. The graph tracks problem identification to closure and not the frequency at which a problem is occurring.

A design issue tracking system works well for small businesses and simple products. It is easy to set up and does not require custom-made or costly

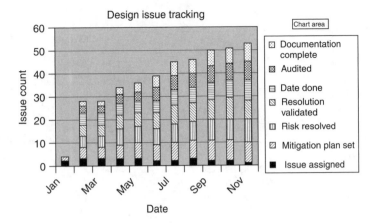

Figure 17.2   Design issue tracking chart

software to manage. This can be set up using any of the common computer software spreadsheet applications, such as Microsoft Excel. Design issue tracking is an alternative method to FRACAS. Whichever system you chose to use (FRACAS or design issue tracking), they need to work in conjunction with an FRB board to resolve problems.

17.2   RELIABILITY GROWTH

FRACAS provides an effective process to identify problems quickly and monitor the progress made toward resolution. However, FRACAS does not indicate how well the product is performing against the reliability goals that were set in the concept phase. Early in the concept phase the reliability goal was set, that is, the product will have a 10,000-h MTBF. After the product is designed and manufactured you need to have some level of assurance that the product will meet its reliability goals. To determine how well the product is performing against the reliability goal requires reliability growth. An example of reliability growth is shown in Figure 17.3. The graph shows the current cumulative and rolling average product reliability for the product. Product improvements that result from the FRACAS activity should be observable in the reliability growth chart. The improvement will be less noticeable in a cumulative MTBF graph where there is a lot of runtime accumulated compared to the short runtime of recent products with the latest improvements. This problem is easily solved by a 13-week rolling average reliability growth curve. The rolling average better illustrates short-term improvements made to the product.

Reliability growth works in conjunction with the FRACAS activity. Typically, when a new product is released, the initial reliability is lower because of the problems identified in FRACAS that are, as yet, unresolved. After each problem is resolved, the product reliability increases because another failure mechanism

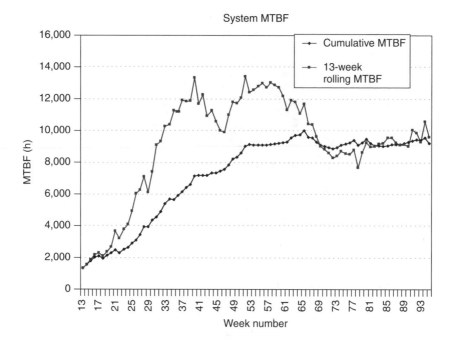

Figure 17.3   Reliability growth chart

has been designed out. This increasing product reliability (for new products that are produced) is tracked using reliability growth. But how do you know if the progress to improve product reliability is acceptable? Are the efforts being made to improve product reliability making a difference to product reliability? Are the reliability improvements being made quick enough to meet the business needs?

The reliability growth curve displays the effect that FARCAS is having on improving product reliability. This gives a snapshot of the current product reliability along with the rate at which it has improved. By tracking reliability growth, a decision can be made regarding the effectiveness of past reliability efforts. However, it is important to determine if the improvements in reliability are occurring too slowly to meet the business needs. Reliability growth can also be used to estimate what the future reliability of the product will be. Knowing this, a decision can be made regarding whether the product design is maturing at a rate fast enough to meet the business needs. An example of this is shown in Figure 17.4.

The future reliability growth can be estimated using a Duane curve. The Duane curve provides an estimate of the time required to achieve the reliability goal based on the current rate of reliability growth. The reliability growth rate is shown in Figure 17.5. Duane observed that the reliability improved on the basis of cumulative MTBF (θ_c) plotted against the total time, on a log–log

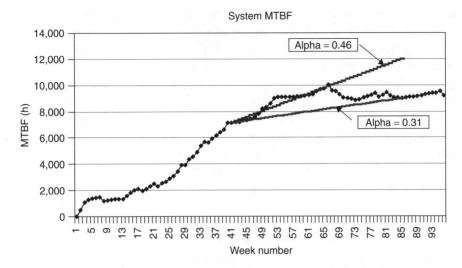

Figure 17.4   Reliability growth chart versus predicted

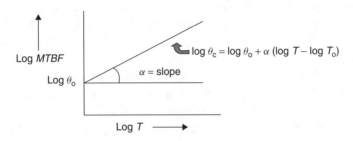

Figure 17.5   Duane curve

scale. This plot could be represented as a straight line using

$$\text{Log } \theta_c = \log \theta_o + \alpha (\log T - \log T_o)$$

where:

$\theta_c = \theta_o (T/T_o)^\alpha$

θ_c = Total cumulative time divided by the total number of failures

θ_o = Observed cumulative MTBF at time T_o

α = Growth rate, $0.1 \leq \alpha \leq 0.6$

T = Expected accumulated product hours, $T > T_o$

T_o = Actual accumulated product hours.

The rate of reliability growth is based on a value "α," which is the slope of the reliability growth curve. If the value of α is closer to 0.1, then the reliability effort to improve product reliability is small and is having little effect. If the value of α is closer to 0.6, then the reliability effort is very ambitious and is having significant effect on improving product reliability. By comparing the desired reliability growth rate to the actual growth rate, a determination can be made regarding the need to make changes in the reliability activity.

The implementation of reliability growth requires assigning a lead person responsible for tracking reliability growth. Often, this is a reliability or quality engineer. However, the skill needed is not at the engineering level, so production personnel are well suited and can be trained for this task. The reliability growth report and the FRACAS report are the primary reports used by the FRB to evaluate progress.

17.3   DESIGN AND PROCESS FMEA

Design FMEAs identify shortcomings in product design and safety and health hazards. When a design is changed because of market needs, added capability, new features, errors, field failures, and so on, a design FMEA can once again protect against design oversights. The entire design doesn't need to have a complete FMEA; usually, just that portion that has been changed. Having a record of the original design FMEA, will greatly speed this process to completion.

Can an FMEA be applied to more than just designs as a tool for improvement? Yes, an FMEA can be applied to manufacturing, assembly, test, receiving inspection, process equipment, and fabrication processes. In fact, the majority of the FMEA activity in this phase is process-related. The process FMEA should be used before implementing any significant changes in the manufacturing process. This ensures that the changes do not impact product quality and reliability. As with design FMEAs, having a record of the original process FMEA will greatly speed this process. This way, any mistakes can be detected and corrected before HASS. This will save time and money.

The FMEA process was described in detail in Chapter 7. The FMEA in this situation is a structured method to study a process that seeks to anticipate and minimize unwanted performance or unexpected failures.

The process is the same for both a design and process FMEA. The major difference is that the participants required will be different. The other major difference in a process FMEA is that it asks the question "what can go wrong with the process?" A process FMEA can also be an effective tool to determine the critical process to monitor for quality control. An FMEA, in this situation investigates the critical processes in the manufacturing process that effect product quality. Next, it determines what process controls need to be in place to prevent rejects and defective products from reaching the customer. The results can then be part of the quality controls used to ensure product quality.

17.3.1 Quality Control Tools

The reliability activities throughout product development ensured that the product was designed for reliability and has ample design and process margin. When a product is designed with sufficient process margin, then controlling quality in manufacturing is all that is needed for success. The quality control tools are designed to achieve this task. There is always a risk of escapes in the manufacturing process where nonconforming products are released for sale. The rate of escapes is related to the first-pass yield of the product. Products with poor yields tend to have higher escape rates than products with high first-pass yields. This is one of the major reasons it is important to have sufficient design and process margin in order to ensure product quality.

In the traditional quality program approach, identify critical processes and continuously monitor and control these processes to an acceptable variability. The focus is on detection and correction of process defects in order to ensure the maximum level of quality. These activities react to process variations that are caught downstream of the manufacturing process. In order for these techniques to be effective, they need to be pushed as far upstream as possible, so that manufacturing escapes are minimized. Ideally, this needs to be done right after any process, where variability is critical. In an assembly process, it can require automated inspection equipment (i.e., X-ray and optical) to detect variation in a critical process such as solder paste deposit, component placement, and solder joint quality. Early detection is the key. Do not rely solely on in-circuit and functional testing to determine product quality. Not only does this minimize the rework cost, and the amount of product affected, but it also minimizes the escape rate of nonconforming products. The technique used to monitor process control in manufacturing is called *Statistical Process Control* (SPC).

There are alternative approaches to quality control. Most notable, is the technique of Quality Function Deployment (QFD). QFD is the process of identifying all factors that might affect the ability of the product to satisfy customer needs and requirements. In essence, identify factors that may affect customer satisfaction. The QFD approach has the advantage of soliciting "voice of the customer" inputs, so issues such as feel and appearance can be considered as well.

SPC

The majority of high-technology products manufactured today are produced to a defined quality standard. The quality standards are achieved through the use of quality control processes. The quality control process incorporates statistical process controls that define upper and lower process control limits. When the process goes out of limits, the production line is stopped, forcing high visibility on the problem and the appropriate people are notified.

Statistical Process Control (SPC) charts use statistical monitoring to control the manufacturing process and maintain tolerances. The benefits are lower

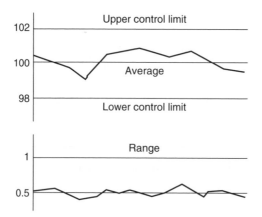

Figure 17.6 Typical SPC chart

production cost, higher yields, less material scrap, reduced warranty cost, and rampability. The process parameters that are chosen to monitor are important. The difficulty comes in being able to identify the critical process parameters that effect product quality. These critical parameters must be controllable.

A generic control chart is shown in Figure 17.6. The control chart has an upper and a lower control limit that must be maintained by the process. The process cannot vary outside these two limits. For this example, a bar is cut to a critical length of 100 cm. A sampling method is selected along with a sampling plan. After a predetermined number of products have been made, a randomly selected sample lot (typically 30) is pulled to verify that the process is in control. The sample lot is measured for cut length and an average length is recoded and plotted on the control chart. The Range chart is the average of the process variations (plus and minus values) around the desired cut length. The range chart indicates that the process is in control as long as the averages of the variations are well within the acceptable control limit. Over time, the process may drift out of control. The control chart ensures that unacceptable process drift is identified, and that the correction actions bring the process back into control.

The time or number of units processed before the process reaches nonconformance is the sample rate of the sampling process. The number of samples and the sampling rate can vary. For stable processes, fewer samples and less frequent sample periods are required for process control. There are techniques to optimize the sampling size and frequency so that inspection cost is minimized. Another benefit to this process is that a minimal skill level is required to run the process control chart. This means that those closest to the process can be responsible for ensuring that the process remains in control and possibly make adjustments when needed. Only when the production personnel can no longer control the process engineers are needed to diagnose and correct the situation.

Control charts can be used to control important parameters that establish the quality of the end product. A few examples are: drill hole depth, output voltage, power out, light level, adhesion, weight, display accuracy in monitors, and so on. The list is determined by what the product quality specifies.

There are many variations of control charts but they *all* must be continuously monitored. The time period between monitoring points varies as a function of the rate of change or process drift before unacceptable errors are encountered. Each SPC data collection system is empirically determined, and each monitoring step in the whole SPC system may well have different monitoring periods. It is best to automate this task. With today's process tools, there is a wide variety of computerized and hand help SPC data collection and charting devices. They can take the drudgery out of the work of data collection, calculation, and charting the results. Applying SPC to the manufacturing process affords a straightforward method of controlling the process for an already well-designed product. For process control, where much greater control and control accuracy is required, the Six Sigma method can be applied.

Implementing SPC into the manufacturing process will require the following:

1. Selecting the key personnel who will learn the SPC skills. They will be the in-house trainers who will work to develop the SPC skills internally. (The quality control manager is a good candidate.)

2. Identifying the key parameters that need to be in control. Which of these should be done first, second, and so on? (Review and Pareto failure reports to reveal nonconformance that is not under control.)

3. Determining the hardware and software that is needed to implement the SPC process with ease and acceptable cost. (There are many companies that produce SPC tools; using the Internet will reveal many.)

4. Defining a "kickoff" day in which everyone involved in the new SPC process will participate.

An example of an SPC process is a wave-soldering machine used in circuit board manufacturing. The product subassembly was an electronic printed circuit board with plated-through holes for the component leads. After the components are "stuffed" onto the board, the leads are trimmed to length and the board is placed on a conveyer system. Then it is slowly passed through a wave-soldering process where the leads are passed over a bath of molten solder. The bath has a device that creates a wave or hump in the bath, so the component leads are immersed for the correct amount of time. Passing through this process the solder cools and the leads remain solidly soldered to the board. The process was not under SPC at the time. One day, the test group found a very high solder defect rate and determined that it was due to the wave-soldering process.

A review of the process, the machine, and the personnel revealed nothing that could be attributed to the cause of the solder defects. However, the problem persisted. Then, the quality manager decided to implement SPC on the wave-soldering process. Soon, he was able to correlate the solder defects to the time of the day. He was still unaware of the cause of the problem but his control charts revealed that the solder temperature took a sudden drop at a certain time of the day, and this could be tracked to the circuit boards that exhibited the defects. He sat on the production line and just observed. Then he saw the new wave-soldering person toss a few ingots of solder into the bath on his way to lunch while the machine was in process. This caused the solder to be low in temperature for a short time. It was later determined that this was the cause of the solder defects. Here, the SPC temperature revealed that the solder bath was driven low enough to cause the defects but not low enough to cause much concern at the wave machine.

Six Sigma

Six Sigma is a process-control method that has gained a lot of acceptance in the past quarter century. It assumes that everything has a normal distribution and that the production process error-detection system can actually measure differences down to one part per million. This accuracy is rarely required. The training required to be an expert in Six Sigma is not trivial. This special skill creates a problem. The talent and knowledge that is already in a company must step aside for the Six Sigma experts, who often can only offer statistical support. Unless the experts are the ones doing the actual corrective actions, there will always be a disconnect between the Six Sigma statisticians and the actual knowledge base in the company.

Establishing the Six Sigma capability within your existing staff will take some months to develop. It is recommended that the simpler SPC process be implemented immediately to get that control mechanism in place, and up and running. After the SPC system is established the time may be right to send key personnel for Six Sigma training. You may find that it may well not be needed for your company, especially in the early stages of reliability improvement.

HASS and HASA

Highly Accelerated Stress Screening (HASS), as described earlier, is a process that is applied to detect unacceptable changes in the manufacturing process or materials that are going into the product. HASS is a more efficient process than the traditional product burn-in commonly used to reduce infant mortality failures. HASA is simply a HASS process that is implemented on a sampling or audit basis.

Traditional product burn-in processes are generally ineffective and costly. Environmental Stress Screens (ESS) is a more effective method to reduce product infant mortality and takes significantly less time to run. However, HASS is, usually, more effective at identifying manufacturing-related defective products than ESS and is less damaging to the product.

All of these "burn-in" techniques take time to run, they require capital cost to set up, and reduce first-pass yield. Convincing management on the need to stress test final products to reduce infant mortality failures is often difficult. The need to ramp product and not impede product ramp is strong. So often, a compromise is required, so both business and quality objectives are achieved. If the product development successfully followed the reliability process, the design should be reliable and have sufficient design margin. If this turns out to be the case, then HASA will turn out to be the most economical end effective method to ensure product quality is maintained.

If the reliability process during product development was cut short, skipped, or omitted, it is likely some form of 100% ESS or HASS testing will be required to weed out defective products. This testing will continue to be required until design changes bring the product failure rate to an acceptable level. The payback from a well-implemented reliability program is seen early in manufacturing, because costly burn-in testing quickly changes to an auditing (HASA) process. There is a significant cost savings from being able to quickly switch to HASA, and not screening every product built.

These tools are used to effect control, monitoring, and the identification of an unacceptable process before nonconformance becomes part of the end product. In addition, there are tools that afford continuous improvement.

References

FMEA

1. *Recommended Failure Modes and Affects Analysis (FMEA) Practices for Non-Automobile Applications*, SAE (2001).
2. M. Krasich, *Use of Fault Tree Analysis for Evaluation of System Reliability Improvements in Design Phase*, 2000 Proceedings Annual Reliability and Maintainability Symposium (2000).
3. K. Onodera, *Effective Techniques of FMEA at Each Life-Cycle Stage*, 1997 Proceedings Annual Reliability and Maintainability Symposium, IEEE (2000).
4. S. Bednarz, D. Marriot, *Efficient Analysis for FMEA*, 1998 Proceedings Annual Reliability and Maintainability Symposium (1998).
5. M. Kennedy, *Failure Modes and Effects Analysis (FMEA) of Flip-Chip Devices Attached to Printed Wiring Boards (PWB)*, IEEE/CPMT International Manufacturing Technology Symposium, IEEE (1998).
6. R. Whitcomb, M. Riox, *Failure Modes and Effects Analysis (FMEA) System Development in a Semiconductor Manufacturing Environment*, IEEE/SEMI Advanced Semiconductor Manufacturing Conference, IEEE (1994).

7. D. J. Russomanno, R. D. Bonnell, J. B. Bowles, *Functional Reasoning in a Failure Modes and Effects Analysis (FMEA) Expert System*, 1993 Proceedings Annual Reliability and Maintainability Symposium, IEEE (1993).
8. S. Prasad, Improving Manufacturing Reliability in IC Package Assembly Using the FMEA Technique, *IEEE Transactions of Components, Hybrids and Manufacturing Technology*, **14**(3), 452–456 (1991).

Quality

1. S. M. Nassar, R. Barnett, *IBM Personal Systems Group Applications and Results of Reliability and Quality Programs*, 2000 Proceedings Annual Reliability and Maintainability Symposium (2000).
2. D. K. Ward, A Formula for Quality: DFM + PQM = Single Digit PPM, *Advanced Packaging* (June/July, 1999).
3. S. B. Lee, A. Katz, C. Hillman, Getting the Quality and Reliability Terminology Straight, *IEEE Transactions on Components, Packaging, and Manufacturing*, **21**(3), 521–523 (1998).
4. Carolyn Johnson, Before You Apply SPC, Identify Your Problems, *Contract Manufacturing* (May, 1997).
5. C.-H. Mangin, *The DPMO: Measuring Process Performance for World-Class Quality*, SMT (February, 1996).
6. H. L. Oh, A Changing Paradigm in Quality, *IEEE Transactions On Reliability*, **44**(2), 265–270 (1995).
7. T. A. Pearson, P. G. Stein, On-Line SPC for Assembly, *Circuits Assembly*, (October, 1992).
8. G. Kelly, SPC: Another View, *Surface Mount Technology* (October, 1992).
9. P. Gupta, Process Quality Improvement – A Systematic Approach, *Surface Mount Technology* (August, 1992).

Reliability growth

1. J. Donovan, E. Murphy, *Improvements in Reliability-Growth Modeling*, 2001 Proceedings Annual Reliability and Maintainability Symposium, IEEE (2001).
2. L. Edward Demko, *On reliability Growth Testing*, 1995 Proceedings Annual Reliability and Maintainability Symposium, IEEE (1995).
3. H. Crow, P. H. Franklin, N. B. Robbins, *Principles of Successful Reliability Growth Applications*, 1994 Proceedings Annual Reliability and Maintainability Symposium, IEEE (1994).
4. G. J. Gibson, L. H. Crow, *Reliability Fix Effectiveness Factor Estimation*, 1989 Proceedings Annual Reliability and Maintainability Symposium, IEEE (1989).
5. J. C. Wronka, *Tracking of Reliability Growth in Early Development*, 1988 Proceedings Annual Reliability and Maintainability Symposium, IEEE (1988).
6. D. K. Smith, Planning Large Systems Reliability Growth Tests, 1984 Proceedings Annual Reliability and Maintainability Symposium, IEEE (1984).

Burn-in

1. D. R. Conti, J. Van Horn, *Wafer Level Burn-In*, Electronic Components and Technology Conference, IEEE (2000).
2. J. Forster, *Single Chip Test and Burn-In*, Electronic Components and Technology Conference, IEEE (2000).
3. T. Sdudo, *An Overview of MCM/KGD Development Activities in Japan*, Electronic Components and Technology Conference, IEEE (2000).
4. C. F. Hawkins, J. Segura, J. Soden, T. Dellin, Test and Reliability: Partners in IC Manufacturing, Part 2, *IEEE Design & Test of Computers*, IEEE (October–December, 1999).
5. W. Kuo, T. Kim, An Overview of Manufacturing Yield and Reliability Modeling for Semiconductor Products, *Proceedings of the IEEE*, 87(8), 1329–1344 (1999).
6. A. W. Righter, C. F. Hawkins, J. M. Soden, P. Maxwell, *CMOS IC Reliability Indicators and Burn-In Economics*, International Test Conference, IEEE (1988).
7. J. Jordan, M. Pecht, J. Fink, How Burn-In Can Reduce Quality and Reliability, *The International Journal of Microcircuits and Electronic Packaging*, 20(1), 36–40, First Quarter (1997).
8. R. Garcia, *IC Burn-In & Defect detection Study*, (September 19, 1997).
9. T. R. Henry, T. Soo, *Burn-In Elimination of a High Volume Microprocessor Using I_{DDQ}*, International Test Conference, IEEE (1996).
10. T. Furuyama, N. Kushiyama, H. Noji, M. Kataoka, T. Yoshida, S. Doi, H. Ezawa, T. Watanabe, *Wafer Burn-In (WBI) Technology for RAM's*, IEDM 93-639, IEEE (1993).
11. P. Thompson, D. R. Vanoverloop, Mechanical and Electrical Evaluation of a Bumped-Substrate Die-Level Burn-In Carrier, *Transactions On Components, Packaging and Manufacturing Technology*, Part B, 18(2), 264–168, IEEE (1995).
12. T. Bardsley, J. Lisowski, S. Wislon, S. VanAernam, MCM *Burn-In Experience*, MCM '94 Proceedings (1994).
13. Michael Pecht and Pradeep Lall, A Physics-Of-Failure Approach to IC Burn-In, *Advances in Electronic Packaging*, ASME, pp. 917–923 (1992).
14. W. Needham, C. Prunty, E. H. Yeoh, *High Volume Microprocessor Test Escapes an Analysis of Defects our Tests are Missing*, International Test Conference, pp. 25–34.

HASS

1. T. Lecklider, How to Avoid Stress Screening, *Evaluation Engineering*, 36–44 (2001).
2. D. Rahe, *The HASS Development Process*, 2000 Proceedings Annual Reliability and Maintainability Symposium, IEEE (2000).
3. M. Silverman, *HASS Development Method: Screen Development, Change Schedule, and Re-Prove Schedule*, 2000 Proceedings Annual Reliability and Maintainability Symposium, IEEE (2000).
4. D. Rahe, *HASS from Concept to Completion*, Qualmark Corporation.

18

End of Life Phase

Eventually, all products reach the end of life when they can no longer serve the needs of the customer both in overall performance and capability. There are many reasons products suffer in performance. Sometimes the product eventually wears out from use; but more than likely it is because other products can outperform the older units.

Computers that were state of the art two to three years ago are being replaced with computers with significant improvements allowing the user greater productivity. In this case, the older computer may be performing to the original specifications but if the market needs higher capabilities the older unit is rendered obsolete. Perfectly good audio devices and cell phones are continuously being replaced with newer devices that in one way or another outperform their predecessors. This is performance obsolescence through new improvements. The time between new product release and obsolescence is getting shorter and shorter. The time to make profits for a given product getting is narrower and narrower. So early market entry and high reliability are very important for profit capture.

Some products are expected to have a long product use life, that is, automobiles, CAT scanners, consumer electronics, and so on. Reliability concerns can surface if the product starts to wear out before its end of life. When this happens, expect more frequent component failures. There may even come a time when the product cannot be repaired because replacement parts are no longer available. Planning for this eventuality can be a lifesaver.

If a product has a long product life, that is, customer demand is sufficient to keep the product on the market for five years or longer, then there is a good chance that some of the parts required to build the system may no longer be available. These components may be obsolete because next-generation technology replaced them; supplier is no longer in business or demand dropped below a manufacturers minimum requirement. Usually, there will be an advance

---

Improving Product Reliability: Strategies and Implementation.   Mark A. Levin and Ted T. Kalal
© 2003 John Wiley & Sons, Ltd   ISBN: 0-470-85449-9

notice of a parts discontinuance or change; this should be part of any purchasing agreement. Once a part obsolescence is noted, a plan for mitigation is required. Some strategies for dealing with obsolescence are as follows:

Seek an alternative supplier (but often similar suppliers obsolete similar component products for similar reasons).

Look for suppliers who specialize in obtaining and storing these components.

On occasion, a producer will learn that a component has become obsolete and a last buy will not satisfy their needs. (These suppliers who specialize in this market may well be used as an early warning ally to help avoid this problem.)

Determine how many of these soon to be unobtainable parts will be needed to support product life for a last time purchase to meet this need.

Once a part is identified as soon to be obsolete, it should be flagged in as "DISCONTINUED" and "DO NOT USE IN NEW DESIGNS."

As part of the new product development process, ensure that there is a Bill Of Materials (BOM) check step that reviews all BOMs for part obsolescence. This will avoid designing in parts whose obsolescence may not be realized until production.

When replacement parts are found, it is good practice to perform a Highly Accelerated Life Test (HALT) on the product to validate performance to specification.

18.1 PRODUCT TERMINATION

At some point in time, a business is compelled to stop manufacturing a product for sale. In addition, a decision is also required to stop supporting a product. The two decisions do not need to occur at the same time. When a product is being phased out, there are some activities that should take place for proper closure. They are as follows:

1. Removing items no longer needed from the warehouse
   (a) Parts and assemblies
      - Scrapped
      - Sold at discount
      - Reworked for new products
   (b) Literature
      - Schematics
      - Manuals
      - Bills of material
2. Transforming the old manufacturing processes out of production
   (a) Manufacturing lines and/or cells
   (b) Production line inventories

  (c) Test fixtures and equipment

  (d) Processes

3. Cessation of sustaining engineering support.

18.2 PROJECT ASSESSMENT

Finally a review should be done to study the lessons learned throughout the product life cycle. What plans could have been put in place to mitigate problems that occurred that might well happen on the next product? Should additional checks be added to the process? Are the planned life cycles matching what really happened with the product? Should life cycles be reviewed, and how often? There are many items and subsequent actions that can be added to the process to develop smoother product life cycles, and going through an end-of-life review will help ensure better outcomes.

Reference

1. R. Solomon, P. A. Sandborn, M. G. Pecht, *Electronic Part Life Cycle Concepts and Obsolescence Forecasting*, IEEE (2000).

19

Field Service

You do your job well and still there are some failures, both in-house and in the field. Knowing this, you must make this eventuality as painless as possible. Doing so will reduce cost and customer dissatisfaction. Taking repair and maintenance into consideration is part of the product design requirements. By design, you can provide a product that is easily maintained and quickly returned to service.

19.1 DESIGN FOR EASE OF ACCESS

When subassemblies or component parts fail, easy access makes the repair faster and more reliable. Designing for easy access usually adds little to no cost to the product; that is, as long as this aspect of the product design is kept in mind during the design phase.

Typically, a product will have several subassemblies that make up the whole. Designing the system so that these smaller units can be removed and replaced quickly and easily is an accessibility plus. Designs should allow the service technician the ability to get to any assembly that is likely to fail without the need for loosening or removing other assemblies. When this is not the case, the handling of the other subassemblies, twisting cables, removing pulleys, belts, brackets, and so on can lead to unanticipated failures in the future or can extend the service time.

When a failed assembly can be removed and replaced without removing other parts, the reliability after repair will be higher. This is because in removing other assemblies to service a part, there is a risk of damaging unaffected assemblies and the added complexity associated with reassembly. We all have the experience where something is serviced for one problem and soon a new problem surfaces that was probably caused by improper reassembly or adjustments.

Improving Product Reliability: Strategies and Implementation. Mark A. Levin and Ted T. Kalal
© 2003 John Wiley & Sons, Ltd ISBN: 0-470-85449-9

19.2   IDENTIFY HIGH REPLACEMENT ASSEMBLIES (FRUS)

Accept that your product will have some failures. Identify those assemblies that are most likely to need service more often than others. Design the system so that these sections can be easily removed, serviced, and replaced, without having to access other parts of the system. Doing so will make the system service event less painful to you and your customer.

When considering replacement items, be sure to identify those that will need periodic servicing and or replacement, for example, fans, filters, belts, drive wheels, fluids, circuit boards, batteries, and so on. Knowing what will need replacement and making this step easy is part of the design effort. Where applicable, identify those items that should be replaced on schedule.

Power supplies, that have cooling fans, have two failure specifications. The Mean Time Between Failures (MTBF) of the supply may be several hundred thousand hours while the fan in the supply may have only a 20,000- to 80,000-h life expectancy. When the fan fails, the power supply will sense fan rotation stoppage and shut down to prevent failures from overheating. Select power supplies with long fan life, not just high MTBFs. See that these fans can be easily replaced – even in the field. Fans will be less expensive to stock for repairs than power supplies. In cases where you return these power supplies to their manufacturer for service, make fan replacement part of the repair contract to ensure that newly repaired power supplies do not fail soon because the fan is about to expire.

Often, several fans are used in larger systems. In a group of the same fans, each having the same life expectancy, one fan in the group will fail first. It is best to replace the whole group, all at the same time. Lubricant loss is the main cause of fan failure. All of the fans in the group will have a similar environment, so the oil loss from each fan will be very similar. The fan that failed first is an indicator that the others will soon follow. Replace them all at once. After you learn the fan failure rate you can initiate a preemptive replacement schedule. They can be replaced during scheduled maintenance, thus avoiding a failure during use.

Fans, motors, pumps, filters, seals, and so on are components that will need to be replaced periodically, much like the fan belt in an automobile. Some subassemblies are attached to these wearout items. Often, a fan is made part of the whole unit, for example, a power supply. This means that the whole subassembly will need to be replaced when a fan fails. Avoid this, by making the fans a subassembly in itself. Subassemblies with fans usually have many wires to disconnect and later reattach when the replacement unit is obtained. Having the fans in a self-contained unit allows for a simpler wiring removal and attachment because with fans only a few wires are required, even for complex fan systems. Field Replacement Units (FRUs) need to be identified as part of the design and passed on to those who provide these items to the field. Remember, that the Failure Modes and Effects Analysis (FMEA) process can help identify FRUs.

Preemptive service planning can make the total service cost lower. Use failure reports [Failure Reporting, Analysis and Corrective Action System (FRACAS)] to identify those areas that are more predictable with regard to wearout, and install service schedules intended for scheduled down times. Apply usage rates and inventory these items to ensure that there are no outages.

In cases where there is a need to replace filters, consider adding a filter flow-sensing component to the design. This can help your customer to avoid downtime during operation. Again, ensure that the filter can be easily replaced. Add low fluid level indicators to avoid catastrophic failures that could be avoided by a simple oil change or a fluid topping off step.

19.3   WEAROUT REPLACEMENT

Some wear items need special accommodations. Incandescent lamps need sockets as do some relays, contactors, circuit breakers, and so on. Sockets can be added to these components to facilitate fast service. However, adding connectors (sockets) of any sort lowers reliability. Be prudent when adding connectors.

19.4   PREEMPTIVE SERVICING

When servicing your own product, consider replacing some parts or components that have shown predictable wearout times. An electrolytic capacitor will lose the fluid in the capacitor much like a fan loses oil from evaporation. The hotter the environment, the sooner this will happen. Check from FRACAS, if this is an issue and replace these parts as part of the repair process. Consider using components that either can last longer in this environment (move from 85 °C to 105 °C electrolytic capacitors because they last approximately four times longer at elevated temperatures). Rotating devices like fans can be affected by how they are mounted (ball bearing devices are more reliable with a vertical shaft position than sleeve bearing devices, however they are much more susceptible to shock). The key is to use FRACAS to identify wearout items that have less than expected product life.

Some sophisticated systems provide monitoring of critical components, that is, X-ray tubes in CAT scanners, the number of contact insertion/removal cycles in large circuit board arrays, the number of times recording tapes/disks are rerecorded, and so on. Some of these top-end systems actually connect to the factory service group via a modem or the Internet, to preemptively order replacement parts for servicing, again during scheduled maintenance. These systems ensure that the service personnel and the replacement hardware arrive at the same time for efficient servicing. This maintains a very high availability.

When examining current field replacement items, it may become apparent that some wear items need replacement too frequently. Here, you can make the change to a more reliable component before the design is finalized. Fixing this problem early in the product development cycle is less costly than doing it after the product is already in the field.

19.5 SERVICING TOOLS

When a system needs an adjustment, see that the design takes into consideration where the adjusting tool will have to go to make the adjustment. Make it easy to insert this tool to make the adjustment. Be aware of the environment in which the system will be. Will there be adequate lighting or will the covers of the system block light needed for proper servicing? Are there adjustments that are read on a scale that is poorly illuminated? Make this area more visible in your design.

The selection of service tools is important too. Design to accommodate tools that are inexpensive and readily available. If your product sales are worldwide, consider tool availability in foreign countries. Even though slotted and Phillips type screwdrivers are common worldwide, avoid them because they can leave small shards caused by scraping the screw head during tightening. These metal icicles can get into places and cause other failures and damage. A slotted-hex-head machine screw is a good alternative. Torx screws are even better and even last longer as replacement screwdriver bits on the manufacturing floor, but they are not as available worldwide. In some cases, a design requires a special tool. When special tools are needed for your product, ensure that it is readily available through your facilities. Remember, however, special tools are expensive to design, manufacture, and stock. Ordinary tools cost much less.

Larger systems may require the removal of very heavy subassemblies. Human strength sometimes cannot do the job. Larger systems often have several heavy subassemblies, for example, power units, air conditioners, large circuit card assemblies, and so on. Make sure that the design teams interface when there is a need for special tools (again consult with the field service personnel). Work to design one tool for all these tasks, instead of one special tool for each major subassembly. Even consider the more ordinary tool need. Select screw sizes that need only one size screwdriver. This too cuts down the cost and quantity of service tools.

The field service group will have a prescribed set of tools that they carry to service the equipment in the field. Get the list, make copies, and pass it on to every designer in the company. Make it clear that these are already the tools they can consider for field use. New tools can be added only upon acceptance by the field service personnel. Make as part of the design verification process a day when the designers get to remove, adjust, recalibrate, and replace their designs. Have everyone perform this service step. The design group will do a much better tool selection job, even beforehand, because they will be thinking of the tool set when they design the product. New tools add cost, make the toolbox heavier, and are often unnecessary. However, sometimes, special tools are necessary.

When you find that the design team has a need for a special tool, ask why? Review all the tools that are part of the service kits and work with the service personnel to see if the existing set will do the job. If all that fails, a special tool may be necessary.

True, service personnel are proud of their tools and to many, the more tools, the better. But after they have to carry them long distances they tend to rethink this idea. Take the time to audit the toolboxes of the service personnel. See if there are tools in their kits that are not on the prescribed list. Learn why any have been added. You may learn that the assembly where this added tool is used cannot be repaired or removed without it, or it is a better tool than the one the designers specified. Service technicians will inevitably add some tools to their service kit. Looking over what they have added, however, can be very instructive.

19.6   SERVICE LOOPS

Some assemblies will need interconnecting cables. Add some length to these cables where possible to allow for easy removal and replacement. This is called *a service loop*. Locate the connectors for these internal assemblies, so that they can be unplugged before the subassembly is removed and plugged back in after the subassembly is securely attached. This added safety precaution is a valuable service feature. Design so that one person can do the job. Sending two repairmen to a job is more expensive. Where customers have in-house service personnel, this can help lower their costs too.

Design cabinetry so that the service personnel can do the work and use meters and other tools without clumsy handling. Adding a power outlet to the cabinet for test equipment is a plus.

When parts are easily removed for service, the time to return the system back to operation is lowered. This too can be part of the design. Some systems require calibration after servicing. Design subassemblies, so that this is taken into consideration. Perhaps, compartmentalization of the subassemblies can make it, so most servicing requires no new adjustments or calibration procedures. This speeds the servicing and makes the system quickly available for use.

19.7   AVAILABILITY OR REPAIR TIME TURNAROUND

Availability is an important part of a design. Design for quick service periods. Availability can be measured and is a metric that can be used with your reliability efforts. The average time between failures is known as the Mean Time Between Failures (MTBF). The time to return a system to the user is the Mean Time To Repair (MTTR). They mathematically relate in a term known as *Availability*, expressed as a percentage of uptime.

$$\text{Availability} = \text{MTBF}/(\text{MTBF} + \text{MTTR})$$

$$\times\ 100(\text{expressed as a percentage of uptime})$$

The equation shows that the availability is greatest when the MTTR is the lowest. A substantial MTBF can be greatly impacted by a large MTTR. (An automobile that has few failures but requires replacement parts form another country is undesirable.)

19.8   AVOID SYSTEM FAILURE THROUGH REDUNDANCY

Where high reliability is required, there may be a need for redundancy. This is where the designers use extra components in tandem, such that when one fails the others can still handle the load. This is commonly done with power supplies. Having five, 500,000-h MTBF power supplies operating as a group where all are needed means that the combined MTBF is 100,000 h. Adding one more power supply in tandem (6 in all), can extend the real reliability of the supply group to several million hours depending on the time planned to inspect for a failed unit and replacing the failed unit before another unit fails. This is referred to as an *N* + 1 *redundancy*. This can be done with switch and connector contacts, fluid lines, and hard drives. The list is endless.

When the reliability of a component is very high and the probability of failure is extremely low, the need for redundancy or local spare parts can be eliminated. Some high-end automobiles have run flat tires; there is no spare for replacement on the road. These tires can run flat at highway speeds for up to a hundred miles so the driver can get to a service facility where replacement is done.

19.9   RANDOM VERSUS WEAROUT FAILURES

It is important to point out that the failures in the field are driven by *random failures* and by *wearout-type failures*. Wearout can be, to a large degree, accommodated in the service planning by design. The accumulation of data in the FRACAS system can help identify the amount of replacement items needed.

Reference

1. D. S. Steinberg, *Vibration Analysis for Electronic Equipment*, Third Edition, John Wiley & Sons, p. 9, Copyright (2000).

Appendix A

The information in this appendix may contain inaccuracies or typographical errors. Every effort was made to ensure the accuracy of this information at the time of final edit. However, business and web information changes over time and without notice. In addition, the information here does not constitute an endorsement by the authors.

RELIABILITY CONSULTANTS

Name	Description
Patrick O'Connor Engineering Management, Quality, Reliability, Safety: Consultancy and Training 62 Whitney Drive, Stevenage, Hertfordshire SG1 4BJ, UK Tel: +44(0)1438 313048 Fax: +44(0)1438 223443 E-mail: pat@pat-oconnor.co.uk	Pat O'Connor provides consulting and training in the practical aspects of quality, reliability, and safety engineering, emphasizing the effective use of design analysis methods, testing, and management. His teaching is based on his books: "Practical Reliability Engineering," "Test Engineering," and "The Practice of Engineering Management."
Dr. Gregg K. Hobbs, P.E. 4300 W. 100th Ave. Westminster, CO. 80031 USA (303) 465-5988 (303) 469-4353 Fax http://www.hobbsengr.com learn@hobbsengr.com	Hobbs Engineering Corporation specializes in teaching and consulting in accelerated reliability techniques such as HALT and HASS, which were invented by Dr. Hobbs. The corporation offers some twenty courses in classical and accelerated reliability methods.

(*continued overleaf*)

Improving Product Reliability: Strategies and Implementation. Mark A. Levin and Ted T. Kalal
© 2003 John Wiley & Sons, Ltd ISBN: 0-470-85449-9

Name	Description
Dr. Jean-Paul Clech EPSI, Inc. P. O. Box 1522 Montclair, NJ 07042 973-746-3796 JPClech@aol.com	EPSI, Inc., is a reliability engineering firm serving the electronics industry providing cost effective solutions to build in the reliability of electronic packages and circuit board assemblies. EPSI developed the solder reliability solutions model and application. Specialty services include SMT/BGA/Flip-Chip/CSP solder joint reliability assessment, and package thermal stress analysis.
Dr. Wayne Nelson 739 Huntingdon Drive Schenectady, NY 12309 518-346-5138 wnconsult@aol.com	Dr. Wayne Nelson is a leading expert on the analysis of reliability and accelerated test data, on which he consults and teaches. He was elected a fellow of the Inst. of Electrical and Electronics Engineers, the Amer. Soc. for Quality (ASQ) and the Amer. Statistical Assoc. (ASA). He authored 120+ publications and two Wiley books ACCELERATED TESTING and APPLIED LIFE DATA ANALYSIS.
James McLinn Reliability Consultant 10644 Ginseng Lane Hanover, Minn. 55341 763-498-8814 JMRel2@aol.com	Reliability consultant helping companies improve their product development process through reliability paper work analysis, improved development steps, robust testing, and data analysis. CRE, CQE, and CQMgr.
Larry Edson Technical Professional for Advanced Reliability Methods General Motors Corporation And CEO of Larry Edson Consulting 21880 Garfield Road Northville, MI 48167 586-578-3375 (General Motors Technical Center) 248-347-6212 (Consulting)	Larry develops "Quick Learning Cycles" using an optimum blend of Qualitative and Quantitative accelerated testing to rapidly mature products prior to production. Mechanical and Electrical failure mechanisms are his specialty, with vast experience in HALT, HAST, and HASS as well as Extrapolated Accelerated Testing using ALTA.

Name	Description
Steinberg & Associates David Steinberg, Ph.D. 3410 Ridgeford Drive Westlake Village, CA 91361 818-889-3636 E-mail: steinbergelctronic@usa.net	Steinberg & Associates have been involved in the design, analysis, evaluation, and testing of sophisticated electronic equipment for reliable operation in severe sine vibration, random vibration, shock, and acoustic noise. Extensive work has also been done in the thermal analysis for steady state and transient conditions. All work includes finite element analysis on high-speed computers. The work involves the latest state of technology in commercial, industrial, and military applications.
Fred Khorasani, Ph.D. 3201 Quail Lane Morgan Hill, CA 95037-6415 408-779-0035 khorasanif@aol.com	Dr. Khorasani holds a Ph.D. in statistics. He is a productivity consultant to companies and a teacher in reliability analysis and reliability improvement, Design Of Experiments (DOE), Taguchi methods, reducing inspection, process and product development, characterization, optimization, and yield improvement.
Reliant Labs Sales 408-567-6912 Lab 408-567-6901 FAX: 408-850-1852 3350 Thomas Rd Santa Clara, CA 95054 www.reliantlabs.com	Reliant Labs specializes in HALT, HASS development, and production HASS. Other services include infrared thermal imaging, ESS, Temperature Only Testing, MTBF Calculations, FMEA, Fault Tree Analysis (FTA), and On-Site Customer Consulting.
System Effectiveness Associates, Inc. 20 Vernon Street Norwood, MA 02062 Phone: (781) 762-9252 Fax: (781) 769-9422 Email: info@sea-co.com http://www.sea-co.com/index.cfm	If reliability is a customer requirement or company mandate we can help! We can provide you with fast, cost effective reliability *analysis* and *test* services to help you manage your reliability goals. Please visit our web site at sea-co.com for more information.

(*continued overleaf*)

Name	Description
Reliability Analysis Associates, Inc. Ed Walbridge 1440 N. Lakeshore Drive. Suite 30F Chicago, IL 60610 312-274-0542 312-274-0574 Fax Web Page: www.reliabilityanalysis.com E-mail: reliability@nidus.com	Reliability Analysis Associates (RAA), Inc. specializes in recruiting reliability engineers for full-time, permanent positions, and in providing such engineers on a contract basis. We also work with engineers having skills closely related to reliability, for example, Test Engineers and System Safety Engineers.
Reliability Center, Inc. P.O. Box 1421 501 Westover Ave. Hopewell, VA 23860 804-458-0645 804-452-2119 (Fax) info@reliability.com www.reliability.com	Reliability Center, Inc. specializes in helping businesses, industry, government, and healthcare organizations improve reliability in all aspects of their operations. The firm provides consulting services, training programs, and software products to clients using their exclusive Opportunity Analysis/Basic FMEA LEAP™System and Root Cause Analysis PROACT® System.

Note: ESS: Environmental Stress Screening; FMEA: Failure Modes and Effects Analysis; HALT: Highly Accelerated Life Test; HASS: Highly Accelerated Stress Screens; HAST: Highly Accelerated Stress Test; MTBF: Mean Time Between Failures.

EDUCATIONAL AND PROFESSIONAL ORGANIZATIONS

University of Arizona
Aerospace & Mechanical Engineering Dept.
Bldg 16, room 200B
Tucson, AZ. 85721-8191
(602) 621-2495
(602) 621-8191 fax

CALCE Electronic Products and Systems Center at University of Maryland
http://www.calce.umd.edu/

FAA Center for Aviation Systems Reliability at Iowa State University
http://www.cnde.iastate.edu/casr.html

The Maintenance & Reliability Center at The University of Tennessee
506 East Stadium Hall
Knoxville, Tennessee 37996-0750, USA
Phone: (865) 974-9625
Fax: (865) 974-4995
E-mail: mrc@utk.edu
http://www.engr.utk.edu/mrc/

University of Maryland
Building 89, Room 1103
College Park, MD 20742
301-405-5323
301-314-9269 (fax)
mailto:webmaster@calce.umd.edu
Electronic Products & Systems Center
http://www.calce.umd.edu/

American Society for Quality
600 North Plankinton Avenue
Milwaukee, WI 53203 USA
800-248-1946
414-272-1734 fax
help@asq.org email
http://www.asq.org/

IEEE Corporate Office
3 Park Avenue, 17th Floor
New York, New York
10016-5997 USA
Tel: +1 212 419 7900
Fax: +1 212 752 4929

Society of Automotive Engineers
SAE World Headquarters
400 Commonwealth Drive
Warrendale, PA 15096-0001 USA
1-877-606-7323 USA
724/776-4841 outside USA

Society of Reliability Engineers
250 Durham Hall
Virginia Tech
Blacksburg, VA 24061-0118
http://www.sre.org/

PROFESSIONAL SOCIETIES

NASA preferred reliability practices and guidelines for design and test
http://msfcsma3.msfc.nasa.gov/tech/practice/prctindx.html

NASA preferred reliability practices with links to other NASA Reliability and Maintainability sites
http://www.hq.nasa.gov/office/codeq/overvw23.htm

Maryland Metrics
http://www.mdmetric.com/

Reliability Analysis Center (RAC)
http://rac.iitri.org/

Society of Reliability Engineers (SRE) – resource providing education, social contact, and insight to foster understanding of reliability, maintainability, and life testing.
http://www.sre.org/

IEEE Reliability Home
http://www.ewh.ieee.org/soc/rs/

IMAPS – International Microelectronics And Packaging Society
http://www.imaps.org/

Emerald Library Sign-on
http://www.emerald-library.com/cgi-bin/EMRlogin

The Annual R & M Symposium (RAMS)
http://www.rams.org/

RELIABILITY TRAINING CLASSES

Barringer and Associates, Inc.
http://www.barringer1.com/

Dr. Gregg K. Hobbs, P.E.

    4300 W. 100th Ave.
    Westminster, CO. 80031
USA
    (303) 465-5988
    (303) 469-4353 Fax
    http://www.hobbsengr.com/

Reliasoft: Reliability Training Seminar Series
Contact:
    Doug Ogden, VP Corporate Relations
    115 S. Sherwood Village Drive
    Tucson, AZ. 85710
    520-886-0410
    520-886-0399 fax

Offers over 17 different courses in reliability and will do in-house training.

James McLinn
  763-498-8814
  clmclinn@aol.com
  jmrel@aol.com

RAC Education & Training
  201 Mill Street
  Rome Hill, NY 13440-6916
  315-337-0900
  315-337-9932 fax
  Email: rac@iitri.org
  rac.iitri.org
Provides training courses worldwide in many areas of reliability. Can also offer on-site training.

Sales#@Reliasoft.com(email)
http://www.Seminars.ReliaSoft.com

TTI: Technology Training Initiative
Contact:
  Colin Stephens, President
  22 East Los Olivios Street
  Santa Barbara. CA. 93105
  805-682-7171
  805-687-6964 fax
  training@Ttiedu.com(email)
  http://www.ttiedu.com

ENVIRONMENTAL TESTING SERVICES

LR Environmental Equipment Co.
Company sells new and refurbished environmental test chambers, HALT & HAST chambers, industrial oven, vacuum equipment, semiconductor equipment, vibration equipment, temperature chamber, aerospace test equipment, laboratory equipment, and other similar types of equipment. Company will also buy your used equipment. http://www.lre.com/

Contech Research, Inc. (Max Peele)
http://www.contechresearch.com/

Chart Industries, Inc.
http://www.mve-inc.com/applied/

More Independent Research and Analytical Labs
http://www.mwrn.com/product/microscopy/morelabs.htm

Sonoscan
http://www.sonoscan.com/

Storagetek
Longmont, CO 80503
303/661-6332 or 800/348-1458
www.storagetek.com

Environ Laboratories
9725 Girard Ave S
Minneapolis, MN 55431
952/888-7795 or 800/826-3710
www.environlab.com

Cascade Engineering Services
2809 152nd Ave NE, Ste 11,
Redmond, WA 98052
www.cascade-eng.com

Trace laboratories
1150 W Euclid Ave
Palatine, IL 60067
847/934-5300
www.tracelabs.com

Anecto Ltd
Mervue Ind Estate
Galway, Ireland
+353-(0)9175-7404
www.anecto.com

Raytheon Analysis Laboratory
131 Spring St.
Lexington, MA 02421
800-RAL-4787
781-860-3380
781-860-3380 FAX
www.Reliability AnalysisLab.com

HALT FIXTURES

Baughn Engineering
Ph: 909.392.0933 Fax: 909.392.0536
2079-B4 Wright Ave. La Verne, CA 91750
Email: fixtures@baughneng.com.
http://www.baughneng.com/

M/RAD Corporation
71 Pine Street
Woburn, Massachusetts 01801, USA

Tel: (781) 935-5940
Toll-Free (888) 500-9578
Fax: (781) 933-7210
E-mail: inquiries@mradcorp.com
http://www.mradcorp.com/contact.html

HALT TEST CHAMBERS

Envirotronics
3881 N. Greenbrooke S.E.
Grand Rapids, MI 49512 U.S.A.
Phone: 1-800-368-4768
Fax: 1-800-791-7237

Extract from Web Site: Envirotronics is a manufacturer of environmental test chambers for temperature, humidity, vibration, altitude, thermal shock, portable shock chambers, burn-in chambers, custom equipment, fluid chillers, air burn-in systems, SAE dust chambers, combustion air units, automotive test systems, drive-in chambers, conditioned rooms, and forced air ovens. Envirotronics also provides emergency service and repair on all makes, preventive maintenance contracts, rebuild or refurbishment of all makes, and A2LA accredited calibrations. http://www.envirotronics.com/

Photograph courtesy of Envirotronics, Inc.

Photograph courtesy of Chart Industries, Inc.

Chart Industries, Inc.
Phone: 1-888-877-3093
Fax: 1-952-882-5188

Extract from the Web: As the global leader and innovator in the cryogenic value chain, Chart is the only company capable of providing complete, turnkey system solutions consisting of the test chamber, vacuum insulated pipe, and liquid nitrogen storage tank for your test chamber requirements. As a complete system supplier, Chart can help design the entire system to ensure optimum chamber performance. Being a larger company, Chart has the resources to design, develop, and manufacture complete chamber systems including benchtop, HALT/HASS, walk-in, and custom chambers. http://www.mve-inc.com/applied/

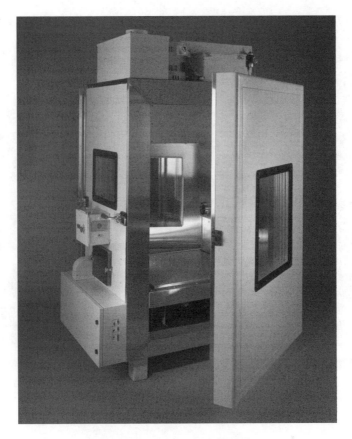

Photograph courtesy of QualMark, Inc.

QualMark
1329 West 121st Ave.
Denver, CO 80234
Phone: (303)254-8800
Fax: (303)254-4372

Extract from web site: QualMark offers a broad range of systems to meet your budget and testing needs. Each of our systems has been specifically designed to provide all the stresses necessary to perform effective HALT and HASS testing. Our systems also are designed to be easy to operate and maintain. http://www.qualmark.com/

Screening Systems, Inc.
7 Argonaut
Aliso Viejo, CA 92656-1423 USA
Tel: (949) 855-1751

Photograph Courtesy of Screening Systems, Inc.

Fax: (949) 588-9910
E-mail: info@scrsys.com

Extract from Web Site: Screening Systems, Inc, QRS series systems, software and services are the complete quality solution for today's business requirements. http://www.scrsys.com/

Thermotron Industries
Holland, Michigan 49423
Voice: (616) 392-1491
Fax: (616) 392-5643
e-mail: info@thermotron.com

Thermotron is one of the industry's premier manufacturers and suppliers of environmental testing equipment. Environmental Test Chambers, Accelerated Stress Test Systems for HALT and HASS, Combined Environment Test facilities, Thermal Shock Chambers, Electrodynamic Shakers, Portable Conditioners, and Test Electronics meet a wide range of industrial testing standards

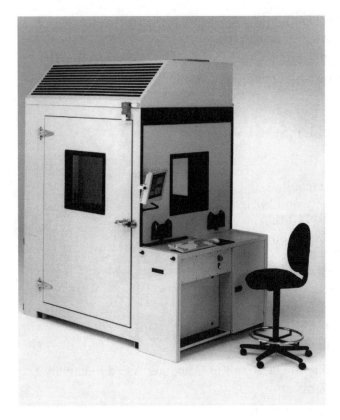

Photograph courtesy of Thermotron, Inc.

and specifications. Whether testing small products such as optical components, integrated circuits, electromechanical devices, flexible circuitry, miniature
electrical components, or large products like automobiles, satellites, missiles,
mainframe computers or telecommunications networking gear, Thermotron
can provide properly designed environmental test solutions to help you predict
and improve quality and reliability. We have systems integration capabilities
and experience to take testing solutions as far as your application demands.
Our technical staff will support you with intelligent designs that improve
testing efficiency: from customized fixturing and sophisticated material handling systems to complete functional electronics and complimentary software.
http://www.Thermotron.com/

RELIABILITY WEB SITES

Barringer & Associates, Inc. Links to Other Reliability Sites
http://www.barringer1.com/links.htm

1. **Adams Six Sigma**
http://www.adamssixsigma.com/

2. **Physics Of Failure Homepage**
http://amsaa-web.arl.mil/rad/pofpage.htm (check link)

RELIABILITY SOFTWARE

ReliaSoft
ReliaSoft Plaza, Suite 103
115 South Sherwood Village Drive
Tucson, AZ 85710 USA
Phone: 520-886-0366
Fax: 520-886-0399
Toll free: 888-886-0410
e-mail: ReliaSoft@ReliaSoft.com
http://www.ReliaSoft.com/

Software Reliability Laboratory
http://www.bsr.uwaterloo.ca/

BQR Reliability Engineering Ltd Computer Aided Reliability Engineering
http://www.bqr.com/

Relex Software Corporation
540 Pellis Road
Greensburg, PA 15601 USA
Phone: 724-836-8800
Fax: 724-836-8844
http://www.relexsoftware.com/

Item Software (USA) Inc.
2190 Towne Centre Place
Suite 314
Anaheim, CA. 92806
Phone: 714-935-2900
Fax: 714-935-2911 e-mail: itemusa@itemsoft.com
http://www.itemsoft.com

Reliass
Cams Hall
Fareham, Hampshire
PO16 8AB
United Kingdom
+44 1329 227 448
+44 1329 227 449 fax

RELIABILITY SEMINARS & CONFERENCES

RAC(Reliability Analysis Center)

Reliasoft: Reliability Training Seminar Series
Contact:
  Doug Ogden, VP Corporate
  Relations
  115 S. Sherwood Village Drive
  Tucson, AZ. 85710
  520-886-0410
  520-886-0399 fax
  Sales#@Reliasoft.com(e-mail)
  http://www.Seminars.ReliaSoft.com

ISTFA: International Symposium for Testing and Failure analysis
Contact:
  Lee Knauss, Publicity Chair
  Neocera
  100000 Virginia Manor Road
  Beltville, MD. 20705
  301-210-1010
  301-210-1042 fax
  lknauss@neocera.com (e-mail)
  http://www.asm-intl.org/istfa

Int'l Symposium on the Physics and Failure Analysis of Integrated Circuits:
Contact:
  IPFA Secretariat
  Kent Ridge Post Office
  P.O. Box 1129
  Singapore 911105
  65-743-2523
  65-746-1095 fax
  ipfa@pacific.net.sg (e-mail)
  http://www.ewh.ieee.org/reg/10/ipfa/

RAMS:

Contact:
  Dr. John English, General Chair
  University of Arkansas
  Department of Industrial
  Engineering
  4207 Bell Engineering Center
  Fayetteville, AK. 72701
  479-575-6029
  chair@rams.org (e-mail)
  http://www.rams.org

Reliability Engineering and Management Institute:
Contact:
  Prof. Dimitri. B. Kececioglu,
  PH.D. PE Dept.
  University of Arkansas
  Aerospace and Mechanical
  Engineering Dept.
  1130 N. Mountain Ave.
  Building No. 119, Room N 517
  P.O. Box 210119
  Tucson, AZ. 85721-0119
  520-621-6210
  520-621-8191 fax
  dimitri@u.arizona.edu(e-mail)
  http://www.u.arizona.edu/~dimitri/

Reliability Testing Institute:

Contact:
  Prof. Dimitri. B. Kececioglu,
  PH.D. PE Dept.
  University of Arkansas
  Aerospace and Mechanical
  Engineering Dept.
  1130 N. Mountain Ave.
  Building No. 119, Room N 517
  P.O. Box 210119
  Tucson, AZ. 85721-0119
  520-621-6210
  520-621-8191 fax
  dimitri@u.arizona.edu(e-mail)
  http://www.u.arizona.edu/
    ~dimitri/

Int'l Symposium on the Testing and Failure Analysis:

Contact:
  AMS International
  Materials Park, OH.
  44073-0002
  440-338-5151
  440-338-4634 fax
  shapowa@asminternational.org
  (e-mail)
  http://www.asminternational.org

Int'l Symposium on the Physics and Failure Analysis of Integrated Circuits:

Contact:
  IPFA Secretariat
  Kent Ridge Post Office
  P.O. Box 1129
  Singapore 911105
  65-743-2523
  65-746-1095 fax
  ipfa@pacific.net.sg (e-mail)
  http://www.ewh.ieee.org/reg/10/ipfa/

Int'l Reliability Physics Symposium:

Contact:
  Eric Snyder
  Sandia Technologies, Inc.
  Albuquerque, NM 87109
  505-872-0011
  505-872-0022 fax
  Eric_Snyder@irps.org(email)
  http://www.irps.org/

Appendix B

MTBF, FIT, AND PPM CONVERSIONS

One of the most often used numbers in reliability is the Mean Time Between Failures (MTBF) number. MTBF represents the average time one can expect a device to operate without failing. There is no assurance that the consumer will realize this failure free time period because the MTBF is a statistical average. In fact, if a consumer experiences a failure, the likelihood of an additional failure is the same before the failure occurs as it is after the failure is repaired (Table B.1).

For example:

If the MTBF $= 8,760$ h
Then on average a unit will fail every 8,760 h or once a year.
1 year $= 356$ days $\times 24$ h/day $= 8,760$

Viewed another way:

If there are 10,000 of these systems in the field, then the manufacturer can expect 10,000 failures every year (for a repairable system), and if the product is nonrepairable system, there will be about 6,700 failures. This is covered in greater detail in the next section.

The Failures In Time (FIT) rate is defined as the failures in time per billion hours.

It is easy to convert between Mean Time Between Failures (MTBF), Failures In Time (FIT), and Parts Per Million (PPM) rates.

Mean Time Between Failure (MTBF)

There is a lot of confusion about the term MTBF. When the layperson hears that a device has an MTBF of 10,000 h, they often think that this means that this device will not have a failure for at least 10,000 h. This is not the case. What this means is that for a group or fleet of systems with an MTBF of 10,000 h the *average rate of failure* will be 10,000 h. Some of the units in this larger group will actually have a failure rate of the stated MTBF rate, while

---

Improving Product Reliability: Strategies and Implementation. Mark A. Levin and Ted T. Kalal
© 2003 John Wiley & Sons, Ltd ISBN: 0-470-85449-9

Table B.1 Conversion Tables for FIT to MTBF & PPM

FIT	MTBF	PPM	FIT	MTBF	PPM
1	1,000,000,000	9	200,000	5,000	1,752,000
2	500,000,000	18	300,000	3,333	2,628,000
3	333,333,333	26	400,000	2,500	3,504,000
4	250,000,000	35	500,000	2,000	4,380,000
5	200,000,000	44	600,000	1,667	5,256,000
6	166,666,667	53	700,000	1,429	6,132,000
7	142,857,143	61	800,000	1,250	7,008,000
8	125,000,000	70	900,000	1,111	7,884,000
9	111,111,111	79	1,000,000	1,000	8,760,000
10	100,000,000	88	1,100,000	909	9,636,000
20	50,000,000	175	1,200,000	833	10,512,000
30	33,333,333	263	1,300,000	769	11,388,000
40	25,000,000	350	1,400,000	714	12,264,000
50	20,000,000	438	1,500,000	667	13,140,000
60	16,666,667	526	1,600,000	625	14,016,000
70	14,285,714	613	1,700,000	588	14,892,000
80	12,500,000	701	1,800,000	556	15,768,000
90	11,111,111	788	1,900,000	526	16,644,000
100	10,000,000	876	2,000,000	500	17,520,000
200	5,000,000	1,752	3,000,000	333	26,280,000
300	3,333,333	2,628	4,000,000	250	35,040,000
400	2,500,000	3,504	5,000,000	200	43,800,000
500	2,000,000	4,380	6,000,000	167	52,560,000
600	1,666,667	5,256	7,000,000	143	61,320,000
700	1,428,571	6,132	8,000,000	125	70,080,000
800	1,250,000	7,008	9,000,000	111	78,840,000
900	1,111,111	7,884	10,000,000	100	87,600,000
1,000	1,000,000	8,760	20,000,000	50.0	175,200,000
2,000	500,000	17,520	30,000,000	33.3	262,800,000
3,000	333,333	26,280	40,000,000	25.0	350,400,000
4,000	250,000	35,040	50,000,000	20.0	438,000,000
5,000	200,000	43,800	60,000,000	16.7	525,600,000
6,000	166,667	52,560	70,000,000	14.3	613,200,000
7,000	142,857	61,320	80,000,000	12.5	700,800,000
8,000	125,000	70,080	90,000,000	11.1	788,400,000
9,000	111,111	78,840	100,000,000	10.0	876,000,000
10,000	100,000	87,600	200,000,000	5.0	1,752,000,000
20,000	50,000	175,200	300,000,000	3.3	2,628,000,000
30,000	33,333	262,800	400,000,000	2.5	3,504,000,000
40,000	25,000	350,400	500,000,000	2.0	4,380,000,000
50,000	20,000	438,000	600,000,000	1.7	5,256,000,000
60,000	16,667	525,600	700,000,000	1.4	6,132,000,000
70,000	14,286	613,200	800,000,000	1.3	7,008,000,000
80,000	12,500	700,800	900,000,000	1.1	7,884,000,000
90,000	11,111	788,400	1,000,000,000	1.0	8,760,000,000
100,000	10,000	876,000			

(*continued overleaf*)

Table B.1   (*continued*)

MTBF	FIT	PPM	MTBF	FIT	PPM
1	1,000,000,000	8,760,000,000	200,000	5,000	43,800
2	500,000,000	4,380,000,000	300,000	3,333	29,200
3	333,333,333	2,920,000,000	400,000	2,500	21,900
4	250,000,000	2,190,000,000	500,000	2,000	17,520
5	200,000,000	1,752,000,000	600,000	1,667	14,600
6	166,666,667	1,460,000,000	700,000	1,429	12,514
7	142,857,143	1,251,428,571	800,000	1,250	10,950
8	125,000,000	1,095,000,000	900,000	1,111	9,733
9	111,111,111	973,333,333	1,000,000	1,000	8,760
10	100,000,000	876,000,000	1,100,000	909	7,964
20	50,000,000	438,000,000	1,200,000	833	7,300
30	33,333,333	292,000,000	1,300,000	769	6,738
40	25,000,000	219,000,000	1,400,000	714	6,257
50	20,000,000	175,200,000	1,500,000	667	5,840
60	16,666,667	146,000,000	1,600,000	625	5,475
70	14,285,714	125,142,857	1,700,000	588	5,153
80	12,500,000	109,500,000	1,800,000	556	4,867
90	11,111,111	97,333,333	1,900,000	526	4,611
100	10,000,000	87,600,000	2,000,000	500	4,380
200	5,000,000	43,800,000	3,000,000	333	2,920
300	3,333,333	29,200,000	4,000,000	250	2,190
400	2,500,000	21,900,000	5,000,000	200	1,752
500	2,000,000	17,520,000	6,000,000	167	1,460
600	1,666,667	14,600,000	7,000,000	143	1,251
700	1,428,571	12,514,286	8,000,000	125	1,095
800	1,250,000	10,950,000	9,000,000	111	973
900	1,111,111	9,733,333	10,000,000	100	876
1,000	1,000,000	8,760,000	20,000,000	50.0	438
2,000	500,000	4,380,000	30,000,000	33.3	292
3,000	333,333	2,920,000	40,000,000	25.0	219
4,000	250,000	2,190,000	50,000,000	20.0	175
5,000	200,000	1,752,000	60,000,000	16.7	146
6,000	166,667	1,460,000	70,000,000	14.3	125
7,000	142,857	1,251,429	80,000,000	12.5	110
8,000	125,000	1,095,000	90,000,000	11.1	97
9,000	111,111	973,333	100,000,000	10.0	88
10,000	100,000	876,000	200,000,000	5.0	44
20,000	50,000	438,000	300,000,000	3.3	29
30,000	33,333	292,000	400,000,000	2.5	22
40,000	25,000	219,000	500,000,000	2.0	18
50,000	20,000	175,200	600,000,000	1.7	15
60,000	16,667	146,000	700,000,000	1.4	13
70,000	14,286	125,143	800,000,000	1.3	11
80,000	12,500	109,500	900,000,000	1.1	10
90,000	11,111	97,333	1,000,000,000	1.0	9
100,000	10,000	87,600			

Table B.1 (continued)

PPM	MTBF	FIT	PPM	MTBF	FIT
1	8,760,000,000	0.1	200,000	43,800	22,831
2	4,380,000,000	0.2	300,000	29,200	34,247
3	2,920,000,000	0.3	400,000	21,900	45,662
4	2,190,000,000	0.5	500,000	17,520	57,078
5	1,752,000,000	0.6	600,000	14,600	68,493
6	1,460,000,000	0.7	700,000	12,514	79,909
7	1,251,428,571	0.8	800,000	10,950	91,324
8	1,095,000,000	0.9	900,000	9,733	102,740
9	973,333,333	1.0	1,000,000	8,760	114,155
10	876,000,000	1.1	1,100,000	7,964	125,571
20	438,000,000	2.3	1,200,000	7,300	136,986
30	292,000,000	3.4	1,300,000	6,738	148,402
40	219,000,000	4.6	1,400,000	6,257	159,817
50	175,200,000	5.7	1,500,000	5,840	171,233
60	146,000,000	6.8	1,600,000	5,475	182,648
70	125,142,857	8.0	1,700,000	5,153	194,064
80	109,500,000	9.1	1,800,000	4,867	205,479
90	97,333,333	10.3	1,900,000	4,611	216,895
100	87,600,000	11.4	2,000,000	4,380	228,311
200	43,800,000	22.8	3,000,000	2,920	342,466
300	29,200,000	34.2	4,000,000	2,190	456,621
400	21,900,000	45.7	5,000,000	1,752	570,776
500	17,520,000	57.1	6,000,000	1,460	684,932
600	14,600,000	68.5	7,000,000	1,251	799,087
700	12,514,286	79.9	8,000,000	1,095	913,242
800	10,950,000	91.3	9,000,000	973	1,027,397
900	9,733,333	103	10,000,000	876	1,141,553
1,000	8,760,000	114	20,000,000	438	2,283,105
2,000	4,380,000	228	30,000,000	292	3,424,658
3,000	2,920,000	342	40,000,000	219	4,566,210
4,000	2,190,000	457	50,000,000	175	5,707,763
5,000	1,752,000	571	60,000,000	146	6,849,315
6,000	1,460,000	685	70,000,000	125	7,990,868
7,000	1,251,429	799	80,000,000	110	9,132,420
8,000	1,095,000	913	90,000,000	97	10,273,973
9,000	973,333	1,027	100,000,000	88	11,415,525
10,000	876,000	1,142	200,000,000	44	22,831,050
20,000	438,000	2,283	300,000,000	29	34,246,575
30,000	292,000	3,425	400,000,000	22	45,662,100
40,000	219,000	4,566	500,000,000	18	57,077,626
50,000	175,200	5,708	600,000,000	15	68,493,151
60,000	146,000	6,849	700,000,000	13	79,908,676
70,000	125,143	7,991	800,000,000	11	91,324,201
80,000	109,500	9,132	900,000,000	10	102,739,726
90,000	97,333	10,274	1,000,000,000	9	114,155,251
100,000	87,600	11,416			

some will fail sooner, and some later. It is understood that with a population of units the average or mean failure rate will be the stated MTBF rate.

Reliability Defined: The probability that a product will operate satisfactorily for a required amount of time under stated conditions to perform the function for which it was designed.

Taking an example of 100 units that have a 1,000-h MTBF; let's find out more about how many failures there will be, how many units will fail before the stated 1,000 failure rate, how many after, and how many units will have more than one failure.

The rate of failure is exponential. Here the expression is:

$$R(t) = N\varepsilon^{-\lambda t} \tag{B.1}$$

the number still surviving without a failure.

N is the number of units shipped; we will use 100.
$\varepsilon = 2.718$ (or the natural logarithm),
λ is the constant failure rate (in failures per million hours).
$t = 1,000\,h$ (for this example).

FIT is sometimes used in place of λ, but it is smaller by three orders of magnitude, or one failure per billion hours of operation. It is read as "Failures In a Thousand million". Therefore 1,000 λ is one FIT.

$$\lambda = 1/\text{MTBF} \tag{B.2}$$

λ and MTBF are inversely related MTBF is Mean Time Between Failures.
So (B.1) becomes:
$$R(t) = N\varepsilon^{-(t)/(\text{MTBF})} \tag{B.3}$$

Example: Let $t = 1,000\,h$

$\text{MTBF} = 1,000\,h$

$N = 100$ new VCRs or TV sets, or any other type of system

$R(t) = 100 \times \varepsilon^{-(1000\,h)/(1000\,h \text{ between failures})}$

$= 100 \times 2.71^{-1000/1000}$

$= 100 \times 2.71^{-1}$

$= 100 \times 0.37$

$R(t) = 37$ Units "STILL WORKING WITHOUT A FAILURE"

This also means that *63 units had failures.* But in 1,000 h shouldn't all 100 units have had a failure? No; but there still were 100 failures!

At first it seems impossible that there were 100 failures and 37 units were still working; but the answer is that of the 63 units that had failures, some had more than one failure. Some had two or there or even more failures. That's where the total of 100 failures comes from. The only way this could happen is when one unit fails, it is quickly repaired and placed back into service. Even after one failure, as soon as it is repaired there are 100 units that are operating that all have an MTBF of 1,000 h. Even after 25 or 50 or 63 failures, as soon as that last failure was repaired there were always 100 units operating; all with an MTBF of 1,000 h. This is considered the number of failures in "repairable" systems.

Estimating Field Failures

Suppose a product has a 1,000-h system MTBF. Then λ will be 1/1,000 or 0.001. This means that every 1,000 h a system will have a failure. With 100 systems then there will be 100 failures in those 1,000 h. Remember that these failures will show up in only 63 units; the other 37 units will exhibit no failures during this time period.

So how many of the 63 units had 1, 2, 3, ... or n failures?

The number of units having more than one failure can be determined using:

$$P(n) = [(\lambda^n \times t^n)/n!] \times \varepsilon^{-\lambda t} \tag{B.4}$$

Where:

$P(n)$ is the percent of units exhibiting n failures.
t is the time duration,
n is the number of failures in a single system, (e.g. 1, 2, 3, ... n).

Let's learn how many units will have 1, then 2, then 3, and so on, failures per unit in the group of 63 units that will exhibit these 100 failures.

But first a short refresher in factorials:

Note: 0! Is defined as equaling 1; and 0! is read as "zero factorial." See Table B2 for a list of common factorials.

For zero failures: (this is the group of 37 units that had no failures in 1,000 h.)

$$P(0) = [(0.001^0 \times 1,000^0)/1] \times 2.71^{-(0.001 \times 1000)}$$
$$= [(1 \times 1)]/1 \times 2.71^{-1}$$
$$= 1 \times 0.37$$
$$P(0) = 0.37 \text{ or } 37\%$$

Table B.2 Factorials

n	$n!$	n factorial	The math
0	0!	Zero factorial	Defined as 0
1	1!	One factorial	$1 \times 1 = 1$
2	2!	Two factorial	$1 \times 2 = 2$
3	3!	Three factorial	$1 \times 2 \times 3 = 6$
4	4!	Four factorial	$1 \times 2 \times 3 \times 4 = 24$
5	5!	Five factorial	$1 \times 2 \times 3 \times 4 \times 5 = 120$

So with 100 units there will be 37 units exhibiting zero failures in one MTBF time period.

How many units will have one failure in 1,000 h?

Substitute 1 for n

$$P(1) = [(0.001^1 \times 1,000^1)/1] \times 2.71^{-1}$$

$$P(1) = (1/1) \times 0.37$$

$$P(1) = 0.37, \text{ or } 37\% \text{ will exhibit one failure.}$$

So with 100 units there will be 37 units exhibiting one failure in one MTBF time period (1,000 h).

How many units will have two failures in 1,000 h?

$$P(2) = [(0.001^2 \times 1,000^2)]/2 \times 0.37$$

$$P(2) = (1/2) \times 0.37$$

$$P(2) = 18\%$$

So with 100 units there will be 18 units exhibiting two failures in one MTBF time period (1,000 h).

$$P(3) = 6 \text{ units exhibiting 3 failures in one MTBF.}$$

$$P(4) = 1 \text{ units exhibiting 4 failures in one MTBF.}$$

$$P(5) = \text{may be 1 unit exhibiting 5 failures in one MTBF}$$

$$\text{(numbers are rounded).}$$

A more simple way of finding the percentage of failures encountered in a given time period is:

$$P(f) = \lambda t \tag{B.5}$$

Table B.3 Repairable versus Nonrepairable Systems Still Operating (in MTBF Time Units)

	1 MTBF		2 MTBFs		3 MTBFs	
	# Fails	# Still operating	# Fails	# Still operating	# Fails	# Still operating
Repairable systems	100	100	100	100	100	100
Nonrepairable systems	63	37	86	14	95	5

Find how many will fail in *one hundredth* of an MTBF time period.

$$P(f) = 0.001 \times 1{,}000/100 \, \text{h}$$

$$P(f) = 0.001 \times 10$$

$$P(f) = 0.01 \text{ or } 1\%$$

Using 100 units this means that 1 unit exhibits the very first failure in 10 h. So the time to first failure is 10 h!!!!!

Which one it will be in the 100 units is a mystery, however ....

Interestingly enough one unit will last for 5,000 h before it finally has its first failure.

Note: These failures have been considered where the failure rate was exponential. There are other failure rates that are Weibull, Log Normal, and more.

Comparing Repairable to Nonrepairable Systems

If the system is comprised of nonrepairable systems, the number of failures in one MTBF period is lower.

In repairable systems, when a unit fails it is quickly repaired and placed back into service. If the unit cannot be repaired, then when one unit fails in 100 systems there will be 99 units operating after the first failure. Then 98, 97, and so on until they all eventually fail. In 100 systems that are nonrepairable, there will be 67 units that will fail in one MTBF time period. If the MTBF were 10,000 h, then 2 MTBFs would be 20,000 h. The comparison between repairable systems and nonrepairable systems is shown in Table B.3.

Index

Improving Product Reliability: Strategies and Implementation.   Mark A. Levin and Ted T. Kalal
© 2003 John Wiley & Sons, Ltd   ISBN: 0-470-85449-9